Springer Series in Advanced Manufacturing

Series editor

Duc Truong Pham, University of Birmingham, Birmingham, UK

The **Springer Series in Advanced Manufacturing** includes advanced textbooks, research monographs, edited works and conference proceedings covering all major subjects in the field of advanced manufacturing.

The following is a non-exclusive list of subjects relevant to the series:

1. Manufacturing processes and operations (material processing; assembly; test and inspection; packaging and shipping).
2. Manufacturing product and process design (product design; product data management; product development; manufacturing system planning).
3. Enterprise management (product life cycle management; production planning and control; quality management).

Emphasis will be placed on novel material of topical interest (for example, books on nanomanufacturing) as well as new treatments of more traditional areas.

As advanced manufacturing usually involves extensive use of information and communication technology (ICT), books dealing with advanced ICT tools for advanced manufacturing are also of interest to the Series.

Springer and Professor Pham welcome book ideas from authors. Potential authors who wish to submit a book proposal should contact Anthony Doyle, Executive Editor, Springer, e-mail: anthony.doyle@springer.com.

More information about this series at http://www.springer.com/series/7113

Li Yang · Keng Hsu · Brian Baughman
Donald Godfrey · Francisco Medina
Mamballykalathil Menon
Soeren Wiener

Additive Manufacturing of Metals: The Technology, Materials, Design and Production

Springer

Li Yang
Louisville, KY
USA

Francisco Medina
Knoxville, TN
USA

Keng Hsu
Tempe, AZ
USA

Mamballykalathil Menon
Gilbert, AZ
USA

Brian Baughman
Surprise, AZ
USA

Soeren Wiener
Scottsdale, AZ
USA

Donald Godfrey
Phoenix, AZ
USA

ISSN 1860-5168   ISSN 2196-1735   (electronic)
Springer Series in Advanced Manufacturing
ISBN 978-3-319-85575-2   ISBN 978-3-319-55128-9   (eBook)
DOI 10.1007/978-3-319-55128-9

Contents

Chapter 1
Introduction to Additive Manufacturing

1.1 Brief History of AM Development

The idea of producing a 3 dimensional object layer by layer came about long before the development of ideas around additive manufacturing. The first concept patented can perhaps be traced back to Peacock for his patented laminated horse shoes in 1902. Half a century later in 1952, Kojima demonstrated the benefits of layer manufacturing processes. A number of additional patents and demonstrations took placed in the time period of 60–80 s that further solidified the idea of producing a 3 dimensional object using a layer wide approach and in the meantime set the stage for introduction and development of processes based on this principle to produce physical prototypes.

Emerged as the rapid prototyping system in 1987, the SLA-1 (SLA stands for Stereolithography Apparatus) from 3D systems (Fig. 1.1) marks the first-ever commercialized system in the world. This process is based on a laser-induced photo-polymerization process patented by 3D Systems' founder, Chuck Hall, wherein a UV laser beam is rastered on a vat of photo-polymer resin. 3D prototypes are formed by curing the monomer resin layer by layer while in between each layer the build platform submerges deeper into the resin vat.

As 3D Systems' rapid prototyping machine kept involving, other players in system and materials development in the field gradually surfaces. In 1988, in collaboration with 3D systems, Ciba–Geigy introduced the first generation of acrylate resins which marks the genesis of a large part of currently available photopolymer resins in the market. DuPont, Loctite also entered the field in system development and resin business. Meanwhile in Japan NTT Data CMET and Sony/D-MEC commercialized the "Solid Object Ultraviolet Plotter (SOUP), and Solid Creation System (SCS), respectively. These systems were also based on the same photopolymerization principle. In the same time period, the first epoxy-based

© Springer International Publishing AG 2017
L. Yang et al., *Additive Manufacturing of Metals: The Technology, Materials, Design and Production*, Springer Series in Advanced Manufacturing,
DOI 10.1007/978-3-319-55128-9_1

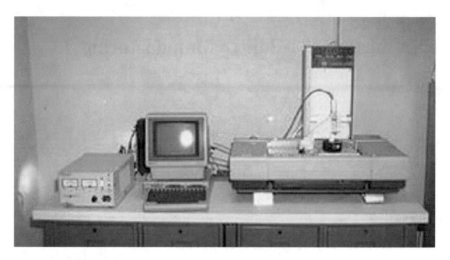

Fig. 1.1 The SLA-1 system by 3D systems launched in 1987

photo-curable resins were introduced by Asahi Denka Kogyo. Epoxy-based resins remain, to this date, another large part of available materials for photopolymerization-based 3D printing methods. In the same time frame, Electro Optical System (EOS) and Quadrax in the European community introduced first stereolithography-based system, while Imperial Chemical Industries introduced the first photopolymer in the visible wavelength range.

Years 1991 and 1993 marked an important milestone of how the current land-scape of additive manufacturing takes shape. Five technologies were commercial-ized the same year: Fused Deposition Model (FDM) from Stratasys, Solid Ground Curing (SGC) from Cubital, and Laminated Object Manufacturing LOM) from Helisys in 1991; soon after that, the Selective Laser Sintering (SLS) from DTM, and the Direct Shell Production Casting (DSPC) from Soligen were introduced. The FDM technology represents a large part of the current landscape of Additive Manufacturing, while the LOM process now has a small market share. The SGC process, though, did not see large commercial success, its operating principle became the forefather of the projection-based SLA system from a number of OEMs and the Continuous Liquid Interface Production (CLIP) technology from Carbon 3D. Both the SLS and the DSPC processes currently occupy a large section of the AM technology market. While the SLS technology has remain largely similar to its initial invention, the DSPC technology has evolved into the systems on which companies such as Ex-one and Voxel Jet base their production machines.

The past two decades marked an accelerated period of AM development. While key existing technologies continued to evolve, new technologies such as the Polyjet materials printing, Laser Engineered Net Shaping, Aerosol Jetting, Ultrasonic Consolidation (a.k.a. ultrasonic additive manufacturing), Selective Laser Melting of metals, and very recently the Continuous Liquid Interface Production technology were demonstrated, developed, and commercialized. Also during the same time frame, existing materials were improved, new materials were demonstrated and commercialized covering polymers, metals, ceramics, composites, foods, and biological materials for a wide range of applications. 3D printed articles are no longer just prototypes. They now can be fully functional end-user parts, assemblies, and even complete systems on scales as small as a "micro-bull" in Fig. 1.2 that is smaller than the diameter of a human hair to functional passenger cars (Fig. 1.3), to dwellable modular homes built by a large gantry system printing concrete, and to the idea of establishing space colonies aided by 3D printing (Fig. 1.4).

A good way of predicting our future direction is to look back and see where we have been. Over the past 30 years it has been evident that the idea of building a 3-dimensional object layer by layer is not only feasible, but it has proven to be potentially something that can shift the entire manufacturing paradigm. In early 2016, every week some form of innovation is being introduced, be it a new technology, a new product, a new material, or a new application. It is not difficult to believe that the next decade with see true breakthrough in the "additive" approach of manufacturing goods in that the phase "design for manufacturing" is no longer needed, but "manufacturing for design" becomes a reality. Total transformation of the manufacturing industry is also in the foreseeable future where the production of a batch of 1,000,000 parts is no longer the sole responsibility of a mass production facility in one or a handful of locations, but hundreds of thousands of "micro factories" in geographically distant locations coproducing in parallel. It is also not absurd to believe that emergency rooms at hospitals can one day be "deploy-able" life-saving mobile stations wherein biological materials are synthesized and 3D printed onto wounds for real-time recovery of injuries.

Fig. 1.2 "Micro-bull" produced by the two pohoton-excitation SLA process by a team of researchers in Japan in 2001. The bull measures 10 microns long. Kawata et al. [40]

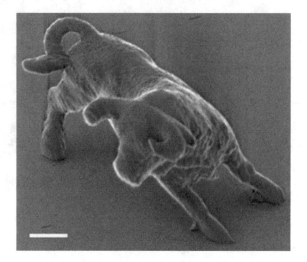

Fig. 1.3 3D printed functional passenger vehicle by Local Motors. *Photo source* speedhunters.com

Fig. 1.4 Artist's depiction of the idea of "Lunar Colonies" utilizing 3D printing technologies. *Photo source* Contour Crafting

1.2 Distinctions and Benefits of Additive Manufacturing

Additive manufacturing process are fundamentally different from "traditional" manufacturing processes such as cutting, forming, casting process. The main difference resides in the fact that in traditional manufacturing processes shaping of materials takes place across the entire physical domain of the desired part whereas in additive manufacturing processes the shaping of material primarily takes place in the formation of the elements (such as voxels, filaments, and layers) which as a whole make up the desire part. The chain of steps in the shaping of elements is implemented in computer-automated environments wherein fabrication of physical 3 dimensional objects from computer-aided design models are accomplished using metallic, polymeric, ceramic, composite, and biological materials.

The distinct process nature of additive manufacturing processes gives rise to a host of advantages over traditional processes. From application perspective, AM offers high degrees of customization and personalization with little impact on manufacturing complexity and cost as the tooling and associated cost component do not exist for AM processes. In the pilot run and low volume production environment, material waste, time and costs associated with materials, inventory are significantly reduced. In addition, geometrically complex, compositionally heterogeneous, and individualized components can be fabricated (for some technologies) while doing so with traditional manufacturing processes can be cost prohibitive. Shown in Fig. 1.5 is an example of an ordered lattice structure not possible by traditional manufacturing processes. The unique characteristics of AM processes fosters innovation as it offers short turn-around time for prototyping and drastically lowered the threshold of production of small-volume end-user products.

Fig. 1.5 3D lattice structure of truncated octahedron unit cells made by the powder bed melting technology. *Photo source* Keng Hsu

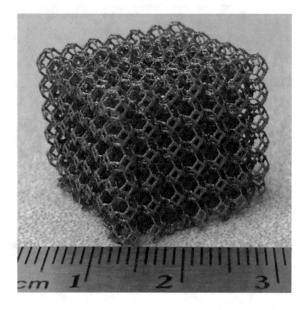

Not bounded by constraints imposed by traditional manufacturing processes, Additive Manufacturing processes may bring about a paradigm shift in the global manufacturing industry as a whole as it enables concepts such as 3D faxing, cloud-based manufacturing, and on-demand end-user location manufacturing.

1.3 Additive Manufacturing Technologies

1.3.1 Material Extrusion

Process Overview

The core process in material extrusion-based AM technologies is the use of an effectively 1D strip (commonly called a road) of material to fill in a 2D space to form one layer. Repeating this process layer by layer one on top of another one allows the formation of a fully defined 3D object. Some of the commercialized technologies based on this process are the Fused Deposition Model (originally patented by Stratasys Inc.) and Fused Filament Fabrication. Shown in Fig. 1.6 are examples of such systems. In these processes a thermoplastic polymer filament is fed through a heated nozzle in which the polymer is heated to above its glass transition or melting point to allow shaping of filament into a "road." These roads fill a layer which, combined with all other layers, forms the 3D object. Once key process characteristic is that the properties of the end product is often anisotropic and highly dependent on the adhesion between each road and all its adjacent roads

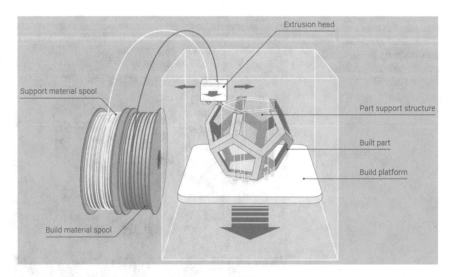

Fig. 1.6 Fused Deposition Model process

in the intra- and inter-layer directions. The reason being that the properties on the interfaces reply on the thermal-activated polymer diffusion process, or reptation. This technology is currently the most wide-spread and has the lowest cost associated with the process and the post-fabrication processes.

As the principle of the extrusion-based process relies on the shaping of a raw material into an extrudate with which an area on a surface is filled to construct one layer of a 3D object, materials with somewhat reversible property changes to allow formation and positioning of the polymer roads. In addition to thermoplastics, thermosets, elastomers, polymer matrix composites, highly viscus liquids, slurry, concrete, and biological media have been shown to be 3D printable using this approach.

The extrusion-based technology is currently the most accessible and flexible in terms of cost and scale of implementation. On the one hand, Do-It-Yourself kits can be sourced easily at very low cost on popular online resellers for hobbyist and educational market users. On the other hand, machines capable of meter-scale engineering structural parts in high fidelity materials such as ULTEM, PEEK are being sold. Continual advancement of this technology are seen in directions such as material availability, process refinement, and improvement.

Process Development

While over the past two decades the FDM technology has seen breakthrough in many areas of development, there are still technological gaps that need to be bridged to bring the FDM technology to the next level of adoption as a manufacturing tool. An example is the low part strength in the build direction of FDM parts. Though the material and process capabilities of this technology has evolved over the years and are now at a point where end-user products can be directly produced, a main property anisotropy issue is still present in FDM parts with optimized build process parameters: the part strength in the direction normal to the build layers is only 10–65% of that in the directions along the filaments with low predictability. This issue places significant design constraints in the growing number of unique engineering applications of FDM-fabricated parts where dynamic loads or multi-direction static loads are present.

So far the FDM process has seen research efforts in mainly five areas: part quality improvement, process improvement, new materials development, materials properties, and applications. Among them three key areas are relevant to the work presented here: (1) material properties (2) process improvement, and (3) part quality improvement. In material properties significant work has been focused in the areas of materials testing, and the use of design of experiments to optimize known process parameters for given part properties. While in-process improvement work has been done to improve support generation process, and to establish numerical simulations of the FDM process, research and development work in FDM part quality improvement has seen progress in accuracy, surface finish, build orientation of parts, and in extension, repeatability. Here we will provide a review of works most relevant to our proposed effort.

In the context of FDM part mechanical behavior, significant amount of work has been put in by various groups that focused on the investigation of effects of FDM process parameters including extruder temperature, raster angle, layer thickness, air gaps in between layers and FDM roads, as well as road widths on the tensile and flexural strengths, elastic behaviors, and residual stresses [1–19]. The conclusions from these studies all point to the same direction: each parameter in the FDM process has different effects on different properties of a part, and that an optimal set of parameters for one property can result in worsening of other properties. An example is that when a minimal dimensional deviation and surface roughness are desired, lower extruder temperatures or active cooling should be used. However, lowering the extruder temperature or the use of active cooling reduces the overall part strength due to less inter-filament and inter-layer bond strength [5].

With the technology evolving for the past few decades mainly under Stratasys, the optimization of process parameters for "best possible" combinations of part dimensional accuracy, surface roughness, and strength is mainly determined in the factory. From end-user's perspective, there is not much that needs to/can be done to improve part qualities. With the original patent expiring in 2009, many low-cost FDM-based 3D printing solutions surfaced and have become an important part of the Additive Manufacturing revolution taking place in the design community. Irrespective of the level of the FDM machine, for part strength isotropy, the "as-built" tensile strengths of parts in the inter-filament/-layer directions fall in the range of 10–65% of that in the direction along the filaments [20]. For the part strength along the directions on a slice/layer, though it also depends on the inter-filament bond strength, it can be remedied by alternating raster angles of adjacent layers (tool path planning) such that in any given direction along the layers the filament-direction strength can contribute to the overall strength of that layer. Here we introduce a "strength isotropy factor" to describe the ratio of the tensile strength of FDM parts in the normal-to-layer direction to the strength in the directions along-the-filament. By this definition, the strength isotropy values would range from 0 to 1 with 0 being the case where there is no strength in the normal-to-layer direction, and 1 being the case where the strengths are the same in both along the layer and across-layer directions. Current FDM parts have a strength isotropy factor ranging from 0.1 to 0.65 with 0.65 being the case with optimized process parameters and a heated build envelope. Almost all work existing in the literature studying part strength properties takes the viewpoint of process parameters and build orientation and their optimization, but only a handful of studies examine the physics of the inter-layer bonding process taking place during FDM and its relation to process inputs.

In a handful of studies, the effect of various process parameters on the bond formation between a "hot" polymer filament and a "cold" existing polymer surface has been investigated. The findings of these studies all indicated that the critical factors that determines the extent of the bond strength between a filament to its adjacent ones lie in the temperatures of the nozzle and the build environment, as well as the heat-transfer processes in the vicinity of the bond site [21, 22]. Of the two temperatures, the build environment temperature has a more significant effect on the bond strength. While it suggests that one could simply increase the build envelope temperature to increase the inter-filament bond strength, the ramification of doing so is that the part dimensional and structural accuracy and tolerances goes way down as the build envelope temperature increase beyond certain points. One team devised a way of heating the entire part surface with hot air during a build process [23]. Though the effect of using hot air was inconclusive due mainly to the approach, the observations very much were in agreement with earlier studies that a critical interface temperature needs to be reached and maintained for a given amount of the time for the bond formation between the filament and the existing surface to go through its three stage of formation: wetting, diffusion, and randomization, much like the reputation model introduced by De Gennes [24] and later adopted by Wool et al. [25].

In 2016 Hsu et al. demonstrated an in-process laser local pre-deposition heating method is reported wherein a near-IR laser supplies thermal energy to a focused spot located on the surface of an existing layer in front of the leading side of the extrusion nozzle as it travels [26]. The principle of this process is shown in Fig. 1.7. As the polymer extrudate comes in contact with the laser-heated region of the surface of the existing layer, the wetting, diffusion, and randomization stages needed to form a strong intermolecular-penetrated bond takes place to a larger extent as compared to deposition processes without local pre-heating. In the results reported here a 50% increase in inter-layer bond strength has been observed. Unlike the current build envelop heating method where the highest temperature used is around half of most polymer's Tg to prevent dimensional and geometrical issues, the laser-based local pre-heating demonstrated in this report is capable of heating extremely locally at only the actual bond site to above its Tg without a negative impact on the part dimension and geometry.

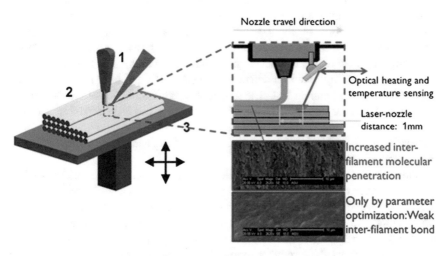

Fig. 1.7 Concept of in-process laser localized pre-deposition heating. This process is demonstrated to be effective in increasing the inter-layer bond strength of FDM parts by more than 50%. Ravi et al. [1]

Fig. 1.8 The flexural strengths of samples at different laser localized pre-deposition heating powers

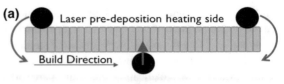

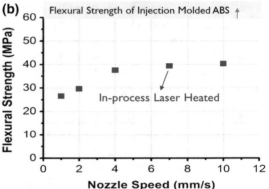

Results were obtained for bending loads required to fracture samples built with laser pre-deposition heating (as shown in Fig. 1.8a). These samples were built at speeds of 1–10 mm/s at the laser intensity of 0.75 W. The flexural strengths of these samples were then calculated from the obtained fracture loads and the geometry of these samples. The flexural strengths of samples built at different nozzle speeds (identical to laser scanning speed) at the same laser heating intensity were plotted in Fig. 1.8b. Our results indicate that, across a range of low laser scanning speeds, the flexural strength of samples built with our laser pre-deposition local heating method increases as the print speed increases; it levels out at above 4 mm/s. This is attributed to the increase in evaporation of material at the high-intensity regions of the laser-illuminated spot as print speed decreases. During FDM, this material evaporation creates a trench into the surface where the incoming extrudate makes contact. If the material flow in the incoming extrudate is not enough to fill the trench, a defect is formed which can later serve as a stress concentrator in the bending test. The defect also reduces the actual cross-sectional surface area to bear the load. In our FDM platform, the laser scanning speed is coupled with the nozzle speed. Therefore, a decrease in nozzle speed results in increased optical energy input into the material surface, causing more material to be evaporated and creating larger defects.

Shown in Fig. 1.9 are the Scanning Electron Micrographs of the fracture surface of the two types of samples (with laser local pre-deposition heated versus without) used in the bending tests. Fracture surfaces are primarily along the surfaces on which cracks propagate during bending tests, and are primarily at the inter-layer interfaces. The results shown here indicate that with laser pre-deposition heating, the fracture surfaces show rougher morphology than those of control samples. We attribute this to the plastic deformation the material adjacent to the interface goes

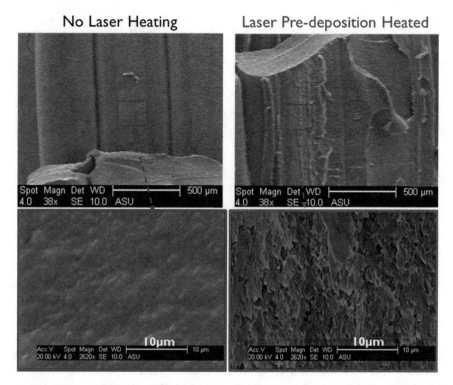

Fig. 1.9 Temperature profiles in two orthogonal planes intersecting the laser spot at two different laser powers

through before the inter-layer interface separates. This indicates that inter-layer interfaces in samples built with the proposed in-process addition show a fracture behavior that is similar to that of the ABS material itself. This change in fracture behavior suggests increased amounts of interpenetrated diffusion across these inter-layer interfaces. This increase in diffusion across the interface allows the inter-layer bond to strengthen as the same interface disappears as a result of interpenetrated diffusion. The fracture surfaces on samples built without the proposed pre-deposition heating method have noticeably smoother morphology. This indicates low crack propagation resistance along these inter-layer interfaces, and that the much smaller degree to which polymer chain interpenetrated diffusion takes place on the inter-layer interfaces in the control samples.

Shown in Fig. 1.7 are the load-deflection relations of samples obtained from the 3-point bending tests. A number of differences in the behaviors between the laser-heated FDM samples and the controls are observed. First, the control samples exhibit a "brittle fracture" behavior where at the end of the linear relation between bending load and deflection, a sharp drop in load is observed that marks the fracture of the samples. On the other hand, the samples with our in-process local pre-deposition heating fail in a ductile behavior where nonlinear load-deflection

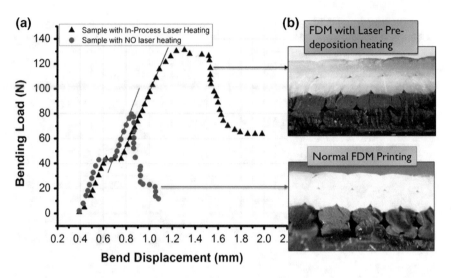

Fig. 1.10 Relationship between bend displacement and bending load

relation is observed. These behaviors agree with our observations of the different morphological appearances on the fracture surfaces of laser heated and control samples. In Fig. 1.9 optical images of the interface between a printed black ABS substrate and a natural ABS layer printed with the laser local heating approach. Discernable differences in the interface geometry are evident in these images between the proposed approach and the intrinsic FDM process. In the laser local pre-heating approach the overall more uniform profile suggest reflow of the substrate surface as it is being heated by the laser beam. It is clear that the proposed approach can not only increase the interface temperature and promote increased amount diffusion, it also reduces the defects in-between roads of filament and between layers (Fig. 1.10).

Shown in Fig. 1.10 are the flexural strengths of samples at different laser localized pre-deposition heating powers. Also plotted here are the flexural strengths of samples built with identical raster, layering, and fill parameters but without laser heating, as well as the range of typical flexural strength values of injection molded ABS. We found that at the chosen set of build parameters, the flexural strength values peaked at 1 W of laser input power. At the peak strength value of 48.2 MPa, a 50% increase in the inter-layer bond strength is observed. It also reaches 80% of flexural strength of injection molded ABS. We attribute this "peak" strength to a local maximum as local heating power increases as a result of two competing mechanisms in the proposed in-process local heating method: (1) the increase in bond strength as the interface temperature increases as a result of increase laser heating power, and (2) the increase in defects created by the evaporation of material at the high-intensity regions of the Gaussian beam. For the second mechanism, it is similar to what we observed in the bond strength at various nozzle speed study wherein as the laser power increases the region of the heated path on the polymer

Fig. 1.11 Relationship between laser intensity and flexural strength

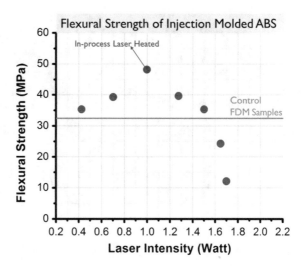

surface where the local temperature reaches, the vaporization point of the material also increases. This causes incomplete fill of each road and generation of defects which can later become stress concentrators during the bending test, giving rise to a decrease in flexural strength as laser power increases. On the other hand, as laser power increases, the inter-layer interfaces see higher temperature. As the temperature increases, the diffusion across the interface can take place to a larger extent and allow higher inter-layer bonds (Fig. 1.11).

The mechanism of the observed increase in inter-layer strength as interface temperature goes up can be explained by the polymer interfacial bond formation model proposed by Yan [22] where the relation between the inter-layer bond strength and the "bond potential" of the interface follows this form:

$$\sigma = \alpha \, e^{1/\varphi}$$

where σ is the bond strength, α is a constant, and φ is the bond potential. The bond potential is a quantity that describes the degree to which an interpenetrated bond can be formed on a polymer–polymer interface. It is a function of interface temperature and time:

$$\varphi = \int_{0}^{\infty} \theta(T) \, e^{-k/T} dt$$

$$\theta(T) = \begin{cases} 1 & T \geq T_c \\ 0 & T < T_c \end{cases}$$

The bond potential is zero (or no bonding) when the temperature of the interface is below the critical temperature of the material. In the case of amorphous ABS, the critical temperature is generally agreed to be the glass transition temperature. When

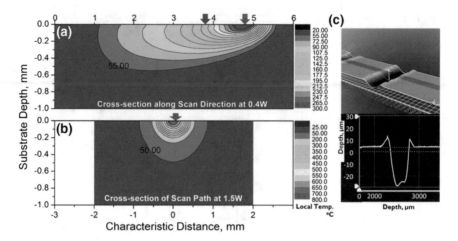

Fig. 1.12 Relationship between characteristic distance and substrate depth

the interface temperature rises above the glass transition, the three stages of wetting, diffusion, and randomization [23] can take place and allow an interpenetrated bond to form. The strength of this bond then becomes a function of temperature, as well as the time duration in which the interface stays above the critical temperature.

Based on our experimental results which agree with the model above, the increase in inter-layer bond strength observed in samples built with our proposed laser local pre-deposition heating is, therefore, attributed to the increase in the bond potential on the inter-layer interfaces as the laser locally heats up the polymer surface prior to a new filament coming into contact with the heated path on the surface. The pre-heated surface region allows for interface temperature to stay above the critical temperature of the material after the contact is made; and for longer periods of time. As a result of increased time and temperature, the diffusion of across the inter-layer interfaces increases, allowing the bond strength to increase.

The spatial temperature distribution in the black ABS substrate as a scanning laser beam traverses across its surface are obtained from a transient heat transfer and thermal model established in commercial finite element modeling package COMSOL. Shown in Fig. 1.12a are the temperature profiles in two orthogonal planes intersecting the laser spot at two different laser powers. As can be seen in Fig. 1.12b, at a 10 mm/s scan speed and 0.4 W laser power, at the center of the illuminated spot the surface of the substrate can be heated to above 250 °C. At 1 mm away from the illuminated spot (where the extruding ABS from the nozzle makes contact with the substrate) the surface temperature, though decreases, remains above 150 °C. Also can be seen is that the heat remains at a shallow depth in that at 0.5 mm below the surface the temperature drops to below 55 °C. The predictions here suggests the localized laser pre-deposition heating is an effective way to locally raise the substrate–filament interface to promote inter-diffusion strengthening without introducing overall heating in the entire workpart, causing mechanical integrity issues during the build. In addition, the reduced temperature gradient across the interfaces during printing can reduce the thermal stresses induced due to the shrinkages of different regions of polymer across layers and roads during printing and cooling (Fig. 1.12).

Figure 1.12b depicts the temperature distribution in the cross-section of scan at a laser power of 1.5 W and a scan speed of 10 mm/s. At this rate of energy input, an affected zone of several hundred microns in width would rise above the thermal decomposition temperature of ABS, resulting in ablation of material. In a set of control experiments performed to verify this, we found that the trench created by laser ablation can result in defects as deep as 30 microns and as wide as 500 microns at a scan rate of 5 mm/s and a laser power of 1 W. Optical profilometry of an example laser ablation profile due to slow speed is shown in Fig. 1.12c.

The correlation between the int,layer bond strength of laser pre-deposition heated FDM builds, measurements of actual surface temperatures and the thermal history need to be obtained. This can be achieved by in-process temperature measurements at the local area illuminated by the laser process beam using through-the-beam optics and IR sensors in the appropriate wavelength range. These are some of the aspects of the proposed approach currently being addressed.

1.3.2   Vat Polymerization

The two key elements of Vat Polymerization-based processes are photo-polymerization resin and a resin exposure system that allow spatial control over polymerization in a vat of monomer resin. There are two primary configurations in this technology: an upright style where the build plate (onto which the desired part is built) is submerged into a vat of resin as the build progresses, and an inverse configuration where resin is contained in a tray and the build plate starts at the bottom of the tray and pulls upwards away from the tray as the build continues. The finished part in the upright configuration is completely submerged in the resin vat, while in the inverse configuration the finished part is completely removed from the resin. Depicted in Fig. 1.13 are the examples of those two configurations. In either configuration, the resin exposure system can be either a rastering laser beam or an image projection-based system.

In the upright configuration, the vat contains all the available resin in the system and is replenished after each build. The build typically starts with the build plate positioned a few hundred microns just below the resin free surface. The exposure system then polymerizes the layer of resin monomer in between the resin surface and the build plate. Once the entire layer is polymerized, the build plate lowers to allow another layer of fresh resin monomer to form between the resin free surface and the previous layer. The process then repeats until the entire 3D objet is completed. Since the viscosity of most resin monomers are typically high, forming a layer of resin monomer of uniform thickness between the build plate in position in the resin and the flow of resin from the edge of an existing layer to fill the layer and

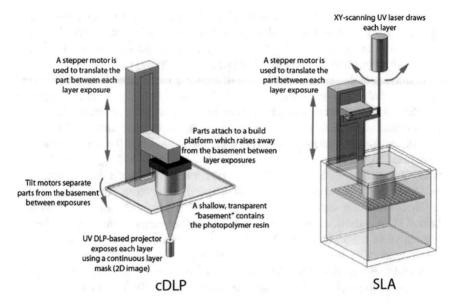

Fig. 1.13 Two configurations of the Stereolithagraphy technology. *Photo source* Wallace et al. [41]

settle back down to form a continuous resin monomer layer with just gravity typically takes a long time. In commercial systems, this issue is typically addressed by one of the two approaches: (1) lowering the build plate by an amount larger than the thickness of one layer to allow reduction of time needed for resin monomer to flow and fill in one layer, and then running a leveling blade across the resin free surface to physically create a "flat" resin surface; (2) a resin dispenser is integrated into the leveling blade to allow simultaneous replenish and leveling in between layers.

In the inverse configuration each layer during the build is defined by the space between the build plate and a solid surface in the vat. Typically the light path is projected upwards into the vat as opposed to the light "shining down" on the resin surface in the upright configuration. Once a layer is completed, separation of the polymerized layer from the solid surface through which the light passes through is needed to allow the resin monomer in the vat to flow into the new space formed and allow the exposure and polymerization of the following layer to continue. One distinct advantage of the inverse configuration is that the thickness and geometry of each layer of resin monomer before polymerization is defined by two stiff solid surfaces using a mechanical compression motion. As a result, the time associated with forming and shaping each monomer layer is reduced as compared with the upright configuration.

Commercial system of both upright and inverse configurations is available, although the inverse configuration is more suitable in an end-user environment.

Currently a major research and development direction in the photopolymerization process-based additive manufacturing approach focuses on the materials development. In addition to elastic and elastomer polymers for structural applications, higher temperature ceramic composites, magnetically, electronically, and thermally functional materials are also of great interests.

1.3.3 Material Jetting

Sharing configuration and system level infrastructure with paper in-jet printing technology, the Material Jetting additive manufacturing process forms each layer of a 3D article by using a single or large arrays of nozzles to deposit droplets of materials on a surface followed by means to cure the deposited material. Shown in Fig. 1.14 is a demonstration of the concept. Initially commercialized by Objet Geometries, the material jetting technology is the fundamental principle of one of the two main technologies commercially offered by Stratasys. On a commercial system, the same materials jetting principle is used for both the model and the support materials required for constructing a 3D component. This technology is widely adopted into prototyping environment as systems based on this technology are often designed for office environments and the build speed, flexibility, and materials availability are ranked high on currently available technologies (Fig. 1.14).

Currently, the materials jetting technology is the only one that offers voxel level material property and color tuning. In Stratasys' "connex" series systems, properties of a given voxel can be "digitally" altered and specified by what is essentially

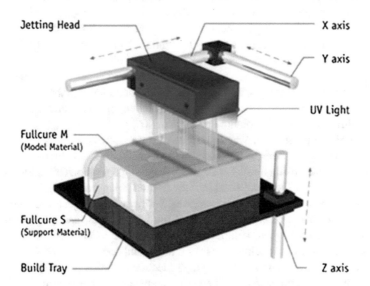

Fig. 1.14 Multi-jet materials jetting technology. *Image source* www.vt.me.edu

mixing of two component model materials of different color or mechanical properties. A typical example is the ability to print hard plastic type of material, rubbery elastomer, and a range of materials of different elastic/elastomer characteristics by mixing the two.

As compared with the Fused Deposition Modeling and the Photopolymerization technologies, the Materials Jetting process can offer higher scalability in productivity, part dimension, and material flexibility owing to the fact that it uses large arrays of nozzles each can act as an individual process channel (as the laser illumination spot in SLA and the extrusion nozzle in FDM). A typical commercial system can have a number of materials jetting nozzles ranging from fifty to several hundreds. Materials available also span a wide range: photopolymers simulating properties of engineering plastics and elastomers, waxes, conductive inks.

1.3.4   Metal Additive Manufacturing Overview

Currently, polymer additive manufacturing, or 3D printing, is accessible and affordable. Systems from $200 Do-it-Yourself kits to half-a-million dollar large scale production printers can be readily acquired. This trend is expected to further develop in the same direction in the near future [26]. Solid metal 3D printing, however, does not see the same development trajectory because of the innate safety concerns and technology and operation costs of existing metal technologies capable of producing solid metal components with densities greater than 95% [27, 28].

Current metal additive manufacturing processes include indirect methods such as Binder Jet processes and Selective Lase Sintering, and direct methods such as Selective Laser Melting, Electron Beam Melting, and Laser Engineered Net Shaping [29]. Indirect methods require post-processing such as Hot Isostatic Pressing to produce parts of density greater than 90% while direct methods can typically produce parts with more than 90% density with optimized process parameters. In indirect methods, metal powders are either partially solid-state sintered together or a low melting point binder is used to bind metal particles together to produce a preform. Post-processing operations such as binder removal, sintering, or liquid metal infiltration are used to obtain greater than 90% build density [27]. The process of Ultrasonic Consolidation was introduced by researchers as a hybrid additive–subtractive process where sheets (or strips) of metal foils are first ultrasonically welded into a stack using a roller sonotrode. A cutting operation (often end-milling) is then used to shape the metal stack into the desired layer shape [30]. By alternating between these welding and cutting processes, 3-dimensional objects are constructed. In UC, the 2D shape of layers is obtained by combining a tape or sheet welding process and the subsequent trimming of welded layer to the desired shape. This process is capable of producing pure metal, alloy, and composite material parts with the use of high power ultrasonics and high mechanical loads [31, 32].

The powder bed process is now entering a stage of technology maturity and is currently the most common metal 3D printing systems for production of engineering components. Current systems of this process use thermal energy to melt and fuse material through manipulation of a meltpool created by laser or electron beam coupled into metal powder as heat. The resulting structures, morphology, and microstructures of printed materials depend highly on the thermal-physical and heat-transfer processes during the micro-welding event [27, 29]. These processes use fine powders as the starting material which can pose health and safety concerns especially when reactive metal powders such as aluminum are used [28]. Because of the heat melt-fusion nature of this technology, the part building process takes place under controlled environment of inert gases or vacuum to prevent excessive oxidation, beam scattering in the case of electron beam melting, and process hazards [34, 28]. Though high-quality metal parts can be produced, a typical powder bed metal system starts at $200,000 with an approximately 800 cm^3 build volume, without taking into account the ancillary equipment and facility required to safely handle and process metal powders. For a system with a build volume practical for production of engineering structural components, half a million to a few millions worth of capital investment alone can be expected.

Though in early stages of development, there has been new processes introduced that can allow 3D printing of metals with greater than 95% density without thermal melt fusion. Hu et al. demonstrated a meniscus confined 3D electrodeposition approach that is capable of producing microscale 3D copper features of more than 99% density [33]. This process showed potential in high-resolution direct printing of fine features with the potential issues of slow deposition rate and limited number of materials available. A number of researchers demonstrated the ability to deposit metals such as copper and silver using a laser-induced chemical reduction process and the ability to have high spatial selectivity and increasing deposition rates [34, 35]. These chemical reduction and deposition-based approaches, however, have common issues in spatial deposition selectivity as the deposited structure increases in height. The process of gas metal arc weld metal 3D printing is a recently developed inexpensive way of building metal components using an automated articulating welding toolhead [36, 37, 38]. This essentially metal welding process allows for a direct writing method where materials are used only when needed (as compared with the powder bed approach where powder is dispersed across the entire build space at each layer). It is, however, based on melt fusion of metals and as a result the thermal physical and thermal mechanical properties are similar to weld joints and to parts made using powder bed melt-fusion processes.

1.3.5   Sheet Lamination

Also recognized as LOM, the sheet lamination process represents one of the two current commercialized technologies that combine additive and subtractive steps to complete a 3 dimensional article. The process takes place by alternating between steps of bonding sheets of materials (polymer, metals, composites, paper, etc.) in a stack through usually heat- or pressure-activated adhesives, and the subsequent

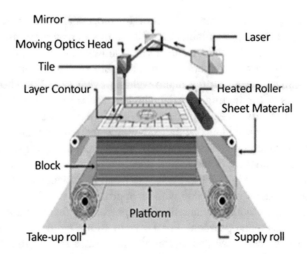

Fig. 1.15 Laminated object manufacturing technology. *Image source* topmaxtech.net

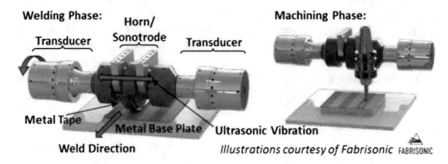

Fig. 1.16 Ultrasonic additive manufacturing technology. *Image source* Fabrisonic

cutting the trimming of the materials stack to the desired contour of a given layer or stacks of layers. The concept of this process is depicted in Fig. 1.15.

In commercial systems, the additive steps of the process are typically accomplished by applying heat, pressure, or both to chemically or mechanically bond sheets together via adhesive that exist on the sheets using a roller. Following the formation of a stack, mechanical cutting and laser cutting are both possible tools to shape each layer or stack of layers. In the case of metal sheet lamination, a process commercialized by Fabrisonic, instead of using adhesives, sheets of metal are ultrasonically welded together to form stacks before the mechanical milling process is brought into shape the layers. The bonds form in between layers, or stacks of layer, are metallurgical through the combination of heat as well as ultrasound-induced diffusion processes across interfaces.

1.3.6   Powder Bed Fusion

There are several variations of the powder bed fusion technology. Although the process capability, conditions, and part characteristics can vary, all variations share the same working principle. As shown in Fig. 1.17 model material in powder form (mean diameter ranges from a few tens of microns to a few hundred microns) is fed into and spread on a build plate driven by a Z-positioning stage by a blade or a wiper mechanism. The space between the surface of the build plate or a finished layer and the bottom edge of this spreading mechanism as it moves across the build area defines the layer thickness and height. The height of a powder layer is typically between a few tens of microns to just below 100 microns in metal systems and between 50um and 150 microns in polymer systems. Layer height is an important factor in the powder bed fusion process, and is carefully selected and calibrated against other build parameters and factors in the system such as energy beam geometry, power, as well as powder particle size and size distribution. Once a layer of powder is formed, an energy beam (laser or electron beam) is focused onto the powder bed and rastered across the powder surface in a pattern to fill the area defined by one slice of the desired 3D model. The raster pattern is also a critical factor and has strong effects on the quality, micro-structure (in the case of metals), and defect structures of material in the completed part. Once a layer of completed, the powder bed process shares similar overall process flow with most other technologies where the build platform drops down and the steps for a single layer repeats to complete the following layer. Note that in the powder bed process the laser exposure is typically adjusted such that a certain depth into the previous layer is also melted to allow full fusion of each layer into the previous. As a result, the properties in the finished part are less directional as compared with those in, for instance, the FDM technology.

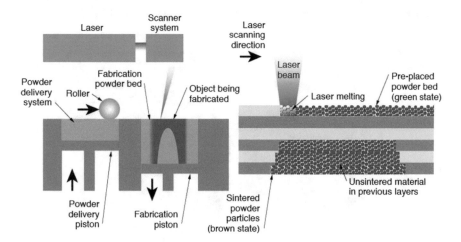

Fig. 1.17 Selective laser melting/sintering technology. *Image source* llnl.gov

The powder bed process is capable of processing a wide range of materials including plastics, elastomers, metals, ceramics, and composites. Depending on the materials used, the desired property, structure, and the limitations of systems used, several variations of process can be implemented. In thermoplastic polymers, the process follows very closely with what was described earlier and nearly fully dense materials can be formed. When metals are used, two variations are possible: SLS and Selective Laser Melting (SLM). In SLS, the powder particles do not fully melt, but are heated by the energy beam to a high enough temperature to allow solid-state diffusion and bonding of powder particles into a porous layer. Alternatively a thermoplastic binder solid is mixed into the model material powder to create a mixture and allow the binder material to be heated by the energy beam during rastering to bind the metal powder particles together. This method is typically used in ceramics when re-melting of material is not possible.

1.3.7   Binder Jetting

The working principles of the binder jet technology resemble the combination of the powder bed process and the materials jetting process. The powder-form model or base material is laid down on a Z-positioning platform to form a layer of bed of powder with uniform thickness across the layer. A print head with single or multiple nozzles similar to those used in the material jetting processes then run pass the powder bed at a given height above the bed and deposit droplets of binder materials to essentially "glue" the powder particles within areas defined by the boundary of a single slice of a 3D model. This layer formation process is repeated as the Z-stage steps down to complete the entire 3 dimensional part. Figure 1.18 shows the working principles of this process. Once completed, the finished part is surrounded by lose powder. To harness the printed part, the Z-platform is raised and the lose powder is removed to reveal the completed print. In the multi-nozzle/-material

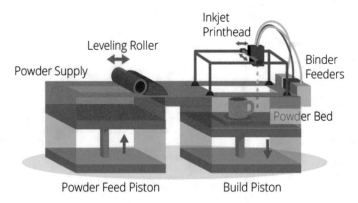

Fig. 1.18 Binder jet binding technology

environment the binder can also be combined with colored ink to provide color printing capabilities.

For polymer applications, oftentimes, the part built by this process is immediately useable without any post-processes. However, in metal or ceramic applications, removal of binder material is required and re-heating of printed part at sintering temperatures is carried out to allow solid-state diffusion to take place among the base material particles to achieve higher strength. At this stage of the post-process, the printed parts are porous. If high bulk density in materials is required, Hot Isostatic Pressing, HIP, or infiltration of a second material is possible. This process has been applied to manufacturing of Injection Molding dies where steel base powder is used to print the molds, and it is followed by a bronze infiltration process to allow full density to be achieved.

1.3.8 Directed Energy Deposition

The Direct Energy Deposition technologies in general refer to processes where the raw material is directed into a spatial location where the energy input and the desired deposition site are co-located. Two processes currently can be categorized under this type: the Laser Engineered Net Shaping (LENS), the Electron Beam Freeform Fabrication (EBF3), and the Wire and Arc Additive Manufacturing (WAAM). These processes are similar in working principles but differ in the energy source used and the form of the raw materials used. Figure 1.19 shows the working principles of these types of processes. In LENS, a laser beam and powder raw material is typically used in an articulating tool head where the powder is injected into a spot on a surface where the laser beam focuses its energy onto. The melt-pool formed at this spot allows the material to be metallurgically added into the existing

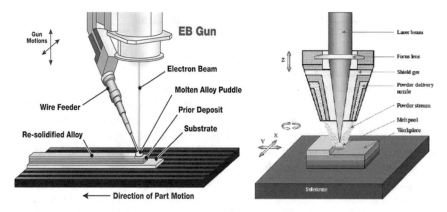

Fig. 1.19 Direct energy deposition processes. *Right* Laser engineered net shaping. *Left* Electron beam free form fabrication. *Source* intechopen.com and sciaky.com

surface and fused into the part being built. By manipulating the spatial locations of the melt-pool, a complete 3D article can be built spot-by-spot, line-by-line, and then layer-by-layer. Following the similar working principle, but different raw material form and energy input, the WAAM process feeds a metal wire into the melt-pool that is produced by the arc struck between the feed wire and the substrate/existing surface. It is essentially an automated Metal Inert Gas (MIG) Welding process in which the weld tool is controlled to follow paths that fill up a 3D part with weld lines. The EBF3 process was first developed by the NASA and is intrinsically a space-compatible technology. It uses the working principles that combine LENS and WAAM in that it uses a wire feed system to introduce materials into a melt-pool generated by an energy source. In EBF3, the energy source used is an electron beam and the build environment is under high vacuum to ensure focusing the operation of the electron beam. The processes in the Direct Energy Deposition category all fall into the category of Direct-Write technology where the material of a 3D part is introduced locally into the part by continuously directing both energy input and material into the same site.

The directed energy deposition has an intrinsic limitation in surface finish and dimensional tolerances on the built part. Though can be implemented as a stand-alone 3D printing process, they are typically configured as hybrid additive–subtractive approaches. In this configuration, it is similar to the Sheet Lamination processes in that the overall process alternates between additive steps and sub-tractive steps where a cutting process is brought in to bring tolerances and surface finish to required range of values. The different processes in this technology are also often times implemented on a 5- or 6-axis articulating system with tool exchange capabilities. It is, therefore, a flexible technology that is well suited for repair of large mechanical or structural components.

1.4   Developmental Additive Manufacturing Technologies

1.4.1   Continuous Liquid Interface Production

CLIP method of 3D printing was first demonstrated in 2014 by Tumbleston et al. where a gas permeable UV-transparent window was used to define layers during the polymerization process. The working principle of this technology is similar to the projection-based vat polymerization processes in the inversed-configuration intro-duced in Sect. 1.3.2. The main difference, which is also set this technology apart from the rest of the photo-polymerization-based technologies in terms of build speed, is that the kinetics of photo-polymerization of resin is coupled with the oxygen-assisted polymerization inhibition, as well as the continuous motion of the built part. The result is that the separation, re-coating, and re-positioning steps in the conventional SLA-type approach are completely avoided. The process rate, there-fore, can be as high as 100 times higher as compared with other types of

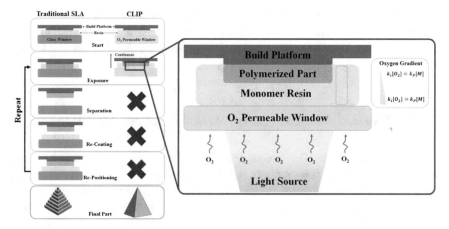

Fig. 1.20 The oxygen permeable window

photopolymerization technologies. As indicated in Fig. 1.20, the oxygen permeable window allows an oxygen concentration gradient to be established on the resin side of the window. This concentration gradient allows the rate of photo-polymerization of resin monomers to follow inverse gradient away from the window. At the build platform speed equal to the polymerization rate (measured in increase in the thickness of the formed layer) of resin, the polymerized part can continue to "grow" as the build platform is continuously pulled away from the window.

At 10 μm or less layer height, this technology produces parts with surface virtually without any stair casing (Fig. 1.21). In the work published a print speed of

Fig. 1.21 Production of parts with surface virtually without any stair casing

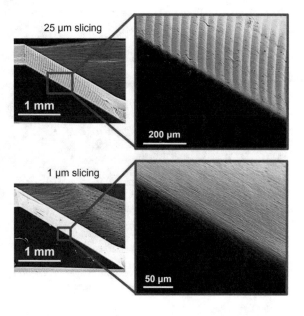

500 mm/h was demonstrated. Being a continuous process, this process rate is limited by the resin cure rates and viscosity, not by layer-wise layer formation. With process tuning and at lower resolution, build speeds of greater than 1000 mm/h can be obtained.

1.4.2   Directed Acoustic Energy Metal Filament Modeling

The process Ultrasonic Filament Modeling (UFM) was first demonstrated by Hsu et al. in 2016 [Patten]. It is currently capable of additively fabricating metal articles of greater than 95% density in ambient conditions at room temperature. The working principle of this process can be thought of as the combination of Wire Bonding and FDM: a solid metal filament is used as the starting material to form a 3 dimensional object via the metallurgical bonding between the filaments and layers. As shown in Fig. 1.22, the mechanics and tooling configuration of UFM is analogous to the FDM process where a heated thermoplastic extruder directly "writes" the tracks and layers that make up the 3D component. In UFM a solid metal filament is guided, shaped, and then metallurgically bonded to the substrate (or the previous layer) as well as the adjacent filaments voxel by voxel using a guide tool

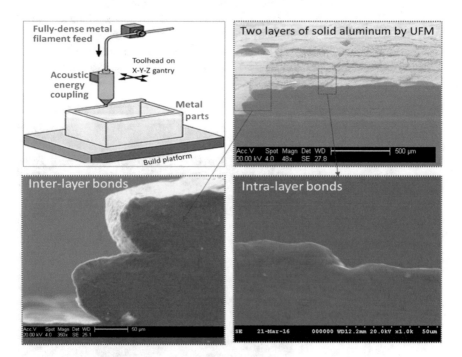

Fig. 1.22 Scanning Electron Microscopy images of a two-layer structure built following the described Ultrasonic Filament Modeling approach

on a positioning system. The key characteristics of this process are that (1) the mechanical stress (and therefore mechanical energy input) required to "shape" the filament into the desired "track" geometry is drastically reduced (<50%) in the presence of applied acoustic energy as compared to the yield strength of the material, (2) the amount of mass transport across the inter-filament and inter-layer interfaces to form the metallurgical bonds observed is more than 10,000 times higher [39] than what Fick's diffusion predicts under the observed conditions, and (3) the temperature rise of the UFM process is nearly negligible (5 °C), a reflection of the high coupling efficiency from acoustic energy input into the required plasticity and mass transport. These unique characteristics enable the Ultrasonic Filament Model process introduced here to be implemented within a desktop 3D printing environment. Additionally, its unique nature of solid metal 3D printing at room temperature enables simultaneous printing of polymers and metals, a materials combination not feasible in melt-fuse-based metal additive manufacturing processes.

In the demonstration work an apparatus based on an ultrasonic energy coupling tool on a 3-axis positioning system was developed to couple acoustic energy into a fully dense aluminum filament, to guide the filament, and to induce the voxel shaping and material fusion required for 3D printing. The ultrasonic vibration source is a piezoelectric crystal-based transducer oscillating in the kHz frequency range. During the UFM process, the tool delivers the acoustic energy to the interfaces between a solid metal filaments and an existing metal substrate as depicted in Fig. 1.22. As acoustic energy is used to shape the metal filament and allow the metallurgical bond on the metal–metal interface to form, the tool steps down the length of the filament to form a "track" of solid metal. These steps are then repeated to form adjacent tracks that make up one layer, followed by repeating track-wise and layer-wise steps to form a 3 dimensional object. In Fig. 1.22 are Scanning Electron Microscopy images of a two-layer structure built following the described Ultrasonic Filament Modeling approach. These images depict filament track shaping as well as metallurgical bonding with an adjacent tracks. In the images shown in the figure, interface regions where native aluminum oxide accumulates can be observed.

In the demonstration work, an L-shaped 3D object and a tensile testing specimen were built following the same procedures. In the case of the L-shaped object, post-process of surface re-finish was used to remove features (less than 200 μm) from the printing process. As shown in Fig. 1.23a through Fig. 1.23d, the object is 5 mm long, 4 mm wide, and approximately 1.5 mm tall with a layer height of 125 microns. The hybrid additive–subtractive mode of the process was also demonstrated. A 16 layer aluminum structure was printed and machined down to an I-shaped object as photographed and shown in Fig. 1.23(c). Preliminary X-ray microtomography results were obtained for the mid-section of the sample, and one representative slice is shown in Fig. 1.23(b). In the micro-CT scans of the UFM printed sample, the inter-layer interfaces are discernable, while no inter-filament interfaces are observed. 99% density is observed. The UFM process introduced here

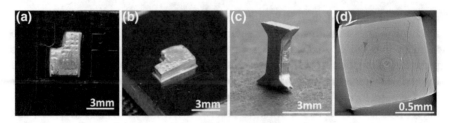

Fig. 1.23 Interface regions where native aluminum oxide accumulates

can be implemented as a direct 3D printing process or a hybrid additive–subtractive fabrication process.

This process allows room-temperature solid metal 3D printing. Both high-resolution IR imaging and thermal couple probing of surface temperatures were used to quantify the temperature rise in UFM. High-speed IR videography captured by a FLIR thermal imaging camera (whose line-of-sight is normal to the cross-section of the aluminum filament and front surface of tungsten carbide tool) shows that the maximum temperature rise for the formation of one voxel is less than 5°. In Fig. 1.24a one frame of thermal video captured during voxel formation shows the spatial temperature distribution in the vicinity of the voxel at the time the maximum temperature is reached. Also presented in Fig. 1.24b is the time evolution of the temperature at the filament–substrate interface. The time evolution of temperature at the critical filament–substrate interface indicates that the fusion of a voxel initiates within 30 ms at the power and force setting used while the shaping of the voxel continues to develop as the voxel process time continues. In Fig. 1.24b, the irradiation of ultrasonic vibrations starts at the 50th millisecond. The relative movements between the two surfaces provide frictional heating that result in the sharp temperature rise. Another 30 ms into the process, the metallurgical bond starts to form and the relative movements between the filament and substrate stops.

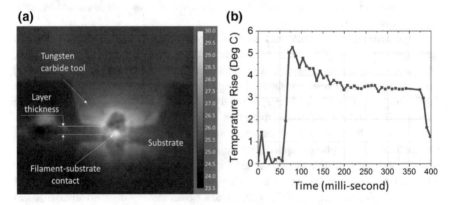

Fig. 1.24 The relative movements between the two surfaces provide frictional heating that results in the sharp temperature rise

This removes the frictional heat source and allows the interface temperature to drop. A maximum temperature rise of 5° is observed. Another feature in the temporal temperature profile is the sharp reduction at 350 microseconds where the ultrasonic vibration stops. This indicates the removal of the second heat source in the process: cyclic plastic strain heating due to the high-frequency cyclic shear deformation in the voxel as it forms.

In the current state the UFM process is capable of printing millimeter-scale pure aluminum and aluminum alloy objects. This process can be implemented as a near-net shape process or a hybrid additive–subtractive process to produce solid metal parts of 95–99% density. During the process a temperature rise of 5° was observed and verified. Dependence of microstructure evolution in the formed metal on input acoustic energy was observed and described. The implication of this is that control of microstructure, and therefore mechanical properties, of printed metal through in-process control of acoustic energy input is possible.

References

1. Ravi A, Deshpande A, Hsu K (2016) An in-process laser localized pre-deposition heating approach to inter-layer bond strengthening in extrusion based polymer additive manufacturing. J Manuf Processes 24(7):179–185
2. Anitha R, Arunachalam S, Radhakrishnan P (2001) Critical parameters influencing the quality of prototypes in fused deposition modelling. J Mater Process Technol 118(1–3):385–388
3. Nancharaiah T, Raju DR, Raju VR (2010) An experimental investigation on surface quality and dimensional accuracy of FDM components. Int J Emerg Technol 1(2):106–111
4. Thrimurthulu K, Pandey PM, Reddy NV (2004) Optimum part deposition orientation in fused deposition modeling. Int J Mach Tools Manuf 44(6):585–594
5. Horvath D, Noorani R, Mendelson M (2007) Improvement of surface roughness on ABS 400 polymer using design of experiments (DOE). Mater Sci Forum 561:2389–2392
6. Wang CC, Lin TW, Hu SS (2007) Optimizing the rapid prototyping process by integrating the Taguchi method with the gray relational analysis. Rapid Prototyp J 13(5):304–315
7. Sood AK, Ohdar R, Mahapatra S (2009) Improving dimensional accuracy of fused deposition modelling processed part using grey Taguchi method. Mater Des 30(10):4243–4252
8. Zhang JW, Peng AH (2012) Process-parameter optimization for fused deposition modeling based on Taguchi method. Adv Mater Res 538:444–447
9. Sahu RK, Mahapatra S, Sood AK (2013) A study on dimensional accuracy of fused deposition modeling (FDM) processed parts using fuzzy logic. J Manuf Sci Prod 13(3):183–197
10. Lee B, Abdullah J, Khan Z (2005) Optimization of rapid prototyping parameters for production of flexible ABS object. J Mater Process Technol 169(1):54–61
11. Laeng J, Khan ZA, Khu SY (2006) Optimizing flexible behavior of bow prototype using Taguchi approach. J Appl Sci 6:622–630
12. Zhang Y, Chou K (2008) A parametric study of part distortions in fused deposition modelling using three-dimensional finite element analysis. Proc Inst Mech Eng Part B 222(8):959–968
13. Nancharaiah T (2011) Optimization of process parameters in FDM process using design of experiments. Int J Emerg Technol 2(1):100–102
14. Kumar GP, Regalla SP (2012) Optimization of support material and build time in fused deposition modeling (FDM). Appl Mech Mater 110:2245–2251

15. Ahn SH, Montero M, Odell D et al (2002) Anisotropic material properties of fused deposition modeling ABS. Rapid Prototyp J 8(4):248–257
16. Ang KC, Leong KF, Chua CK et al (2006) Investigation of the mechanical properties and porosity relationships in fused deposition modelling-fabricated porous structures. Rapid Prototyp J 12(2):100–105
17. Sood AK, Ohdar RK, Mahapatra SS (2010) Parametric appraisal of mechanical property of fused deposition modelling processed parts. Mater Des 31(1):287–295
18. Percoco G, Lavecchia F, Galantucci LM (2012) Compressive properties of FDM rapid prototypes treated with a low cost chemical finishing. Res J Appl Sci Eng Technol 4 (19):3838–3842
19. Rayegani F, Onwubolu GC (2014) Fused deposition modelling (FDM) process parameter prediction and optimization using group method for data handling (GMDH) and differential evolution (DE). Int J Adv Manuf Technol 73(1–4):509–519
20. Masood SH, Mau K, Song WQ (2010) Tensile properties of processed FDM polycarbonate material. Mater Sci Forum 654:2556–2559
21. Ahn SH, Montero M, Odell D, Roundy S, and Wright PK (2000) Anisotropic material properties of fused deposition modeling ABS. Rapid Prototyp 8(4):248
22. Rodriguez JF, Thomas JP, Renaud JE (1999) Tailoring the mechanical properties of fused-deposition manufactured components. Proceedings rapid prototyping and manufacturing '99, vol 3. Society of Manufacturing Engineers, Dearborn, pp 629–643
23. Yan Y, Zhang R, Guodong H, Yuan X (2000) Research on the bonding of material paths in melted extrusion modeling. Mater Des 21:93–99
24. Partain S (2007) Fused Deposition Modeling with localized pre-deposition heating using forced air. Master's Thesis, Montana State University
25. De Gennes PG (1971) Reptation of a polymer chain in the presence of fixed obstacles. J Chem Phys 572
26. Stansbury J, Idacavage M (2016) 3D printing with polymers: challenges among expanding options and opportunities. Dent Mater 32(1):54–64
27. Gu D, Meiners W, Wissenbach K, Poprawe R (2012) Laser additive manufacturing of metallic components: materials, processes and mechanisms. Int Mater Rev 57(3):133–164
28. Wolff I (2016) Breathing Safely Around Metal 3D Printers. Advanced manufacturing.org. 2 Aug 2016
29. Guo N, Leu MC (2013) Additive manufacturing: technology, applications and research needs. Front Mech Eng 8(3):215–243
30. White D (2003) Ultrasonic consolidation of aluminum tooling. Adv Mater Processes 161:64–65
31. Ram G, Robinson C, Yang Y, Stucker B (2007) Use of ultrasonic consolidation for fabrication of multi-material structures. Rapid Prototyp J 13(4):226–235
32. Sriraman M et al (2011) Thermal transients during processing of materials by very high power ultrasonic additive manufacturing. J Mater Process Technol 211:1650–1657
33. Thomas D, Gilbert S (2014) Costs and cost effectiveness of additive manufacturing. NIST Spec Publ 1176
34. Hu J, Yu M (2010) Meniscus-confined three-dimentional electrodeposition for direct writing of wire bonds. Science 329(16):313–316
35. Semenok D (2014) LCLD Laser processing technology for microelectronics printed circuit boards of new generation. Proceedings of the 7th international conference interdisciplinarity in engineering, vol 12, pp 277–282
36. Kochemirovsky VA et al (2014) Laser-induced copper deposition with weak reducing agents. Int J Electrochem Sci 9:644–658
37. Anzalone G et al (2013) A low-cost open-source metal 3D printer. IEEE Access 1:803–810
38. Haselhuhn A et al (2015) In situ formation of substrate release mechanisms for gas metal arc weld metal 3D printing. J Mater Pocessing Technol 226:50–59

39. Haselhuhn A et al (2016) Structure-property relationships of common aluminum weld alloys utilized as feedstock for GMAW-based 3D metal printing. Mater Sci Eng, A 673:511–523
40. Kawata S, Sun HB, Tanaka T, Takada K (2001) Finer features for functional microdevices. Nature, 412:697
41. Wallace J et al (2014) Validating continuous digital light processing additive manufacturing accuracy and tissue engineering utility of a bye-initiator package. Biofabrication, 015003

Chapter 2
Additive Manufacturing Process Chain

A series of steps goes into the process chain required to generate a useful physical part from the concept of the same part using additive manufacturing processes. Depending on the technology and, at times the machines and components, the process chain is mainly made up of six steps:

- Generation of CAD model of the design;
- Conversion of CAD model into AM machine acceptable format;
- CAD model preparation;
- Machine setup;
- Part removal;
- Post-processing.

These steps can be grouped or broken down and can look different from case to case, but overall the process chain from one technology remains similar to that of a different technology. The process chain is also constantly evolving and can change as the existing technologies develop and new technologies surface. In this text, the focus will be on the powder bed metal technology. Therefore, the process chain for this technology will be discussed in details, while other will be roughly mentioned.

2.1 Generation of Computer-Aided Design Model of Design

In any product design process the first step is to imagine and conceptualize the function and appearance of the product. This can take the form of textual descriptions, sketches, to 3-dimensional computer models. In terms of process chain, the first enabler of AM technologies is 3D digital Computer-Aided Design

© Springer International Publishing AG 2017
L. Yang et al., *Additive Manufacturing of Metals: The Technology, Materials, Design and Production*, Springer Series in Advanced Manufacturing,
DOI 10.1007/978-3-319-55128-9_2

(CAD) models where the conceptualized product exist in a "computer" space and the values of its geometry, material, and properties are stored in digital form and are readily retrievable.

In general the AM process chains start with 3D CAD modeling. The process of producing a 3D CAD model from an idea in the designer's mind can take on many forms, but all requires CAD software programs. The details of these programs and the technology behind them is outside of the scope of this text, but these programs are a critical enabler of a designer's ability to generate a 3D CAD model that can serve as the start of an AM process chain. There are a large number of CAD programs with different modeling principles, capabilities, accessibilities, and cost. Some examples includes Autodesk Inventor, Solidworks, Creo, NX, etc. The examples here are, by no means, representatives of CAD technologies, but intended to provide a few key words with which a search on topics on CAD can provide specific information a reader needs.

Once a 3D CAD model is produced, the steps in the AM process chain can take place. Though the process chain typically progresses in one direction that starts with CAD modeling and ends with a finished part or prototype, it is often an iterative process where changes to the CAD model and design are made to reflect feedback from each steps of the process chain. Specific to the metal powder bed technology, critical feedback can come from geometry and property an-isotropy on parts due to build orientation, distortion of part or features due to thermal history of build, issues in generating and removal of support structures, etc. Issues like these may rise in the AM process chain and may call for design changes and revisions. Similar to the topic of "design for manufacturability" in the conventional manufacturing space, design for additive manufacturing is critical and is developing in parallel with the technologies themselves. Chapter 5 is devoted to discussion of design for AM.

2.2   Conversion of CAD Model into AM Machine Acceptable Format

Almost all AM technology available today uses the STereoLithography (STL) file format. Shown in Fig. 2.1 is an example part in its STL format. The STL format of a 3D CAD model captures all surfaces of the 3D model by means of stitching triangles of various sizes on its surfaces. The spatial locations of the vertices of each triangle and the vectors normal to each triangle, when combined, these features allow AM pre-process programs to determine the spatial locations of surfaces of the part in a build envelope, and on which side of the surface is the interior of the part.

Although the STL format has been consider the de facto standard, it has limitations intrinsic to the fact that only geometry information is stored in these files while all other information that a CAD model can contain is eliminated. Information such as unit, color, material, etc. can play critical role in the functionality of the built part is lost through the file translation process. As such it places

Fig. 2.1 An example part in STL format

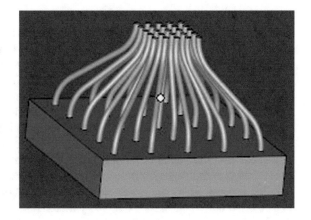

limitations on the functionality of the finished parts. The "AMF" format was developed specifically to address these issues and limitations, and is now the ASTM/ISO standard format. Beyond geometry information, it also contains dimensions, color, material, and additional information is also possible with this file format. Though currently the predominate format of file used by AM systems and supported by CAD modeling programs is still the STL format. An increasing number of CAD program companies, including several major programs, have included support of AMF file formats. Currently, actual use of the information stored in the AMF file is still limited due to the capabilities of current AM systems and the state of current technology development.

2.3 CAD Model Preparation

Once a correct STL file is available, a series of steps is required to generate the information an AM system needs to start the build process. The needed information varies, depending on the technology but in general these steps start with repairing any errors within the STL file. Typical errors can be gaps between surface triangle facets, inverted normal where the "wrong side" of a triangle facet is identified as the interior of the part. Once the errors have been repaired, a proper orientation of the 3D model with respect to the build platform/envelope is then decided. Following the orientation, the geometry, density, geometry of support structures are decided and generated in 3D model space and assigned to the part model. The process then progresses to slicing the 3D model defined by the STL as well the support structure into a given number of layers of a desired height each representing a slice of the part and support models. Within each slice the cross-sectional geometry is kept constant. Once 2D slices are obtained, a "unit" area representing the smallest material placement is then used, combined with a set of strategies, to fill the area within each layer that is enclosed by the surface of the part and the support. For the SLA process, the unit area is related to the laser spot size and its intensity distribution as well as absorption of monomer, and the strategies of filling the enclose area in one layer is the path in which the laser would raster the resin surface. At this point the surface data that are originally in the STL file has been processed and machine specific information to allow placement of the material unit into the desired location in a controlled manner to construct the physical model layer by layer (Fig. 2.2).

Specifically for metal powder bed processes, a STL file is first imported into a software that allows repairing and manipulating of the file, as well as the generation of support, and the slicing of the part and support models. The sliced data are then transferred into the AM system machine for build preparation and the start of the building process. There are a number of software programs that allows these tasks to be carried out, Magics, for example by Materialise is one such software program that is capable of integrating all CAD model preparation steps into one program and generating data files directly accepted by powder bed machine systems. In the sections below additional details of model preparation specific to the powder bed metal process is described.

Fig. 2.2 Support structure generated on the model

2.3.1 STL File Preparation

The CAD model preparation starts with importing an STL, or other compatible file formats, into the pre-process software program (e.g., Magics). Once imported, the dimensions can be modified if needed. Once the model is in desired dimensions, a series of steps is carried out to correct possible errors in the model file. These errors can include missing triangles, inverted or double triangles, transverse triangles, open edges and contours, and shells. Each type of error can cause issues in the building process or result in incorrect parts and geometries. While some errors such as shells and double triangles are non-critical and can sometimes be tolerated, errors such as inverted triangles and open contours can cause critical issue in the building process and needs to be resolved during STL preparation.

2.3.2 Support Generation

For metal powder bed process, the primary function of the "support" structure is to extract heat from the model and to provide anchor surfaces and features to the build plate to avoid warpage due to thermal stresses during and after the build. It does not "support" the part against gravity that causes over hanging or angled features to fail to build.

Generation of support structures in powder bed processes can be accomplished in a few different ways. Also applied to any other AM processes, the first way is to generate the support structures during CAD modeling and design the support to be features of the geometry of the part. Alternatively, the support structures can be generated in the STL preprocess software program. This second approach provides much more flexibility in terms of being able to tailor the structures based on the detailed needs. For example, since the support structure is only used during the build process and is removed during post-process, the amount of material that goes into it needs to be minimized. However, since the primary function of the support is to conduct heat and provide mechanical anchor, a minimum amount of cross-sectional area in the support is needed for it to be functional. Optimization of the volume, geometry, location, and the part-support interface geometry, is important and part dependent. Therefore, carful design of support structure plays a critical role in the success of a build process. Figure 2.2 shows examples of support generated in CAD model space.

2.3.3 Build File Preparation

Once the CAD models of part and support are generated and prepared in the pre-process software program, a slicer program is used to divide the models into layers in the build direction based on the desired layer thickness. For typical metal powder bed systems the layer thickness can be anywhere from 25 microns to close 100 microns. Typical thicknesses used are 25 microns for high resolution builds and 50–75 microns or higher for high-rate applications. Layer thickness is also correlated to the powder dimension and distribution. Ideally the layer thickness would be slightly larger than the mean diameter of powder to achieve high coupling of laser energy input into the absorption, heating and melting of powder, and re-melting of previous layer. At larger thicknesses, scattering of optical energy input may not be sufficient to allow uniform optical energy absorption and heating, resulting in partial melting.

Within each slice, the slicer software also determines the rastering path that the energy beam takes to fully melt the entire layer. How a meltpool created by an optical beam can move in an enclosed area to ensure every part of it is covered is analogous to how an end-milling tool does a pocket milling operation to create an enclosed 2D area with a depth on a surface. Figure 2.3 shows how the use of contour scan coupled with in-fill to fully cover an area. As seen in the illustration, an important factor is the beam diameter, which is captured by the Beam compensation parameter. This value is typically chosen such that sufficient over lapping of adjacent paths occurs and partial melting is avoided.

Other important parameters includes "islands" within each layer, and shifting of scan directions, offset of scan lines from layer to layer as build progresses. The use of islands is important in that the rapid heat generation and cooling in the build part as a result of absorbing a fast-moving optical energy input source (~ 1000 mm/s). Heat input into small "islands" of randomized locations within a layer allows the

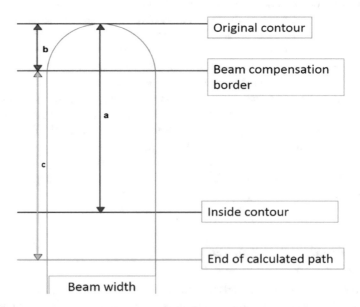

Fig. 2.3 Beam compensation to determine correct beam paths

heat to be evenly introduced spatially, and reduces the chance of deformation and failures caused by uneven warpage during a build. Changing of scan directions and offsetting scan lines from layer to layer allows for uniform distribution of scanline-wise thermal and microstructural anisotropy in the entire 3D space within a build part. It allows for the overall isotropy of material properties.

In addition to rastering strategy, critical parameters that determine energy input into the powder bed to achieve controlled melting are beam power (laser or electron beam), scan speed, and focus move (if the hardware of the systems allows). Overall, these parameters determine the amount of energy incident onto the powder bed per unit time, and directly relates to the heating, melting, and cooling of the material. Each factor, however, also plays a role in different characteristics of the build. High power at faster scan rate produces different thermal history as compared with lower power at slower scan rates, even if the total power input are the same. As a result of the differences in thermal history, the microstructures and properties of parts can differ. In addition, the advective flow of molten material in the meltpool can also have different behaviors when the meltpool travels at different rates, resulting in also differences in the solidification process.

Once the slice information is generated, it is transferred into the interface program that runs on AM systems. The interface program serves as the interface between information of the build and machine controls that carry out the actual build process.

2.4   Machine Setup

Following software preparation steps in the AM process chain, machine preparation is the next step before a build a can start. Machine preparation can roughly be divided into two groups of tasks: machine hardware setup, and process control. Hardware setup entails cleaning of build chamber from previous build, loading of powder material, a routine check of all critical build settings and process controls such as gas pressure, flow rate, oxygen sensors, etc. Details of how each task in this group is carried out can vary from one system to another, but overall once the machine hardware setup is complete, the AM system is ready to accept the build files (slices generated from previous step) and start the build.

The tasks in the process control task group allow an AM system to accept and process the build files, start the build, interrupt the build at any given time if desired or required, and preparing the machine for finished part extraction, and unloading of material. This first task is usually importing and positioning of build parts in the area defined by the build plate. In this step some capabilities of scaling and basic manipulation of build part are usually provided to account for changes needed at this step. Once the physical locations of parts are decided upon, it is followed by a series of steps of defining the (1) build process parameters, (2) material parameters, and (3) Part parameters.

The build process parameter controls machine level parameter that is applied to the entire build. Examples of these parameters include gas injection processes, material recoater motions, and ventilation processes, etc. These parameters define the basic level machine operation to enable a build environment. Material parameters typically control powder dosing behaviors and chamber environment control through inert gas injection. Critical parameter such as "dose factor" determines the amount by which the dose chamber plate rises as compared with the drop in the build chamber build plate. A 100% does factor means that the rise in the dose chamber plate is the same as the drop in the build chamber plate. Typically higher values (150–200%) are desired at the beginning of the process to ensure full coverage of a new layer with powder material. This factor needs to have values greater than 100% because as the powder is melted and fused into a solid layer, the volume occupied by the material in a solid layer form is smaller than that of in the powder form, due to elimination of spaces between powder particles. Therefore, the surface of melted areas is lower than the rest of the powder bed. As the portion of the powder bed surface occupied by build part increases, the dosing factor also needs to increase accordingly. This factor is typically adjustable anytime during a build process as well to provide adjustments to the build process as needed.

Inert gases such as nitrogen or argon are typically used in AM system to control the build chamber environment and maintain low oxygen concentration. Oxygen concentration inside the build chamber is of critical importance to not only the build quality, but also to the success of a build process. Typically the oxygen concentration is maintained below 1–2%. Upper and lower limits of concentrations are often set to allow the gas injection and ventilation system to maintain the oxygen content. Above threshold values, AM systems will shut down the build process. In reactive material (such as aluminum, titanium, etc.) powder bed processes, oxygen content control is particularly important due to safety reasons, and the inert gas injection typically remains on even after the build process ends.

Part parameters are assigned to each and every component/part to be build. Multiple sets of part parameters can be used in the same build on different parts. These parameters are taken into account in the slicing process that takes place in the previous step of the process chain. As a result, the parameters chosen in the slicer have to correspond to the actual parameters selected on the AM system. Once part parameters are selected the building process, the build process starts and is controlled and monitored by the AM system itself. Some feedback and in-process monitoring systems are possible. Most current systems are outfitted with basic in-process diagnostic tools such as melt-pool monitoring where an diagnostic beam coaxial to the process beam monitors the intensity of emission of thermal radiation from the melt-pool and does basic analysis of melt-pool size and radiation intensity spatial distribution. A quality index can be extracted from the results of monitoring to provide an indication of part quality. Another type of in-process feedback tool available to some of the current systems is related to the powder re-coating process. The tool typically takes an optical image of each and every layer and use the reflectivity information within the optical image to determine where a full coating on a competed layer is achieved. In some cases, this value is used to pause the process to prevent failure of builds due to insufficient powder coating and over-heating.

2.5 Build Removal

The build time of the powder bed process depends on a number of factors. Of them, the height of the entire build has the largest effect on the total time. It can take anywhere from minutes to days. Nevertheless, once the build completes, the laser metal powder bed technology allows for immediate unpacking of build chamber and retrieval of finished part, because the process does not maintain the build platform at elevated temperatures (as opposed to laser powder bed for polymers and electron beam-based powder bed processes). The unpacking process typically involves raising the platform in the build chamber and removing loose powder at the same time. The loose powder from one process can be re-used and has to go through a series of sieving steps to remove contaminates and unwanted particulates. Figure 2.4 shows an example of the process. Once the loose powder is removed from the finished part, the build is ready for post-process. The finished parts in metal powder bed AM at this point are welded onto the build plate via support structures. Removal of finished part from the build plate typically involves the use of cutting tools such as band saws, or wire EDM for higher fidelity and flexibility.

Fig. 2.4 SLM part being extracted from build chamber. *Image source* topmaxtech.net

2.6 Post-processing

Depending on AM technology used to create the part, the purpose, and requirements of the finished part, the post-fabrication processes can vary in a wide range. It
can require anything from no post process to several additional steps of processing
to change the surface, dimensions, and/or material properties of the built
part. Shown in Fig. 2.5 is an example of the unique surface features on powder bed
AM parts where partially melted particles are bound to the surfaces of built parts.
These features, along with the weld-lines rastering the melt-pool in different
directions results in a new type of surface finish that is distinct from any existing
manufacturing processes. In metal powder bed AM systems, the minimum required
processing is removal of built part from build plate and the removal of support
structures from the built part. Removal of support structures can be as simple as
manually breaking the supports from the surface of the part, but it can also be a
process that utilizes CNC tools to not only remove the support, but also to achieve
desired surface finish and/or dimension tolerance. In addition, the metal powder bed
AM systems can introduce large amounts of thermal stresses into the built
part. Under these conditions, the support structure serves as the mechanical
"tie-downs" to hold the built part in place and keep it in its intended geometry. If
the supports were to be removed from the part, warpage in the part will occur.
A thermal annealing process can be used to relieve the thermal stresses in the part
before it is removed from the build plate, to prevent part warpage upon removal
from the build plate.

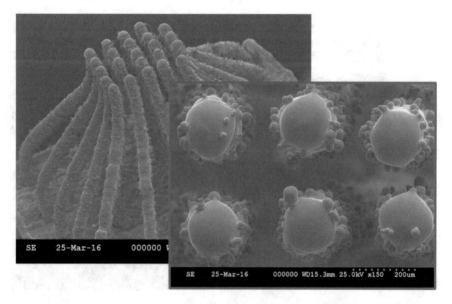

Fig. 2.5 SEM images of SLM parts showing unique surface features unlike any other current
manufacturing processes

Hot Isostatic Pressing, HIP, is a process where a component is subjected to elevated temperature and isostatic pressure in a pressure vessel. At a temperature of over 50% of the melting point of materials and pressures above 100 MPa (can be as high as 300 MPa), the voids and porosity inside a powder bed metal AM part can be greatly reduced. Once processed by HP, the final bulk density can reach more than 95% of the true density of material. Under these extreme conditions of pressure and temperature, material in a component being treated not only undergoes localized plastic deformation, but the processes of creep and solid-state diffusion bonding are allowed to take place. This enables the required shaping and mass transport around internal defects to happen and to "heal" them, increasing the bulk density of the component.

Chapter 3
Microstructure, Mechanical Properties, and Design Considerations for Additive Manufacturing

3.1 Introduction

Microstructure has a direct influence on the mechanical properties of metallic materials. An alloy's microstructure is a result of its inherent chemistry, manufacturing process, and heat treatment. Historically, alloys have been developed and optimized to meet their intended requirements in conjunction with a specific manufacturing process, such as casting, forging, or sheet forming. Additive Manufacturing (AM) is a relatively new process that offers significant benefits for rapid design and implementation and is currently being considered by many industries including aerospace. Many of the already existing aerospace alloys are capable of being processed into viable components by a variety of AM methods. Current expectation is that research in AM processing will lead ultimately to development of unique microstructures that will have a significant impact on the alloy's mechanical properties.

This paper discusses the microstructure and properties of traditional aerospace alloys processed by AM with respect to potential use in gas turbine engine applications. In the past few years, Honeywell has evaluated the AM capability of a wide variety of aerospace alloys. Mechanical property data were generated to determine process capability in comparison to the corresponding cast or wrought versions and to identify potential failure modes in their design and applications. The materials were manufactured in a laboratory environment in the absence of any AM material specifications and controls. Currently, AM lacks an established industrial supply base and the required standards, specifications, and quality system to meet aerospace requirements.

© Springer International Publishing AG 2017
L. Yang et al., *Additive Manufacturing of Metals: The Technology, Materials, Design and Production*, Springer Series in Advanced Manufacturing,
DOI 10.1007/978-3-319-55128-9_3

3.2 Specimen Manufacturing

Specimens of various alloys were manufactured as cylindrical rods or rectangular plates using powder bed laser fusion technology (PBLFT). Three build directions were investigated which were vertical, horizontal and 45° with respect to the horizontal build platform. After building, the specimens remained on the build plates and were subjected to a stress relief treatment to mitigate possible cracking from as-built residual stresses. After stress relief, the blanks were excised from the plates and hot isostatic pressed (HIP). HIP was employed to reduce the occurrence of defects such as pores or possible unbonded regions which could affect fatigue strength. Following HIP, the specimens were heat treated in a manner consistent with their corresponding wrought versions.

Nearly a dozen such alloys have been studied at Honeywell in the past 3 years. Table 3.1 provides a list of materials and the applied post-hip heat treatments. Test samples were machined into standard specimens and tested per ASTM recommended procedures. Some specimens were tested with the additive manufactured as-built surface in-tact in the gage section in order to ascertain its influence on fatigue strength.

3.3 Design Considerations

Component stresses at the temperature of application dictate the proper use of a material. The expected failure mode against which the design has to have an acceptable margin of safety may be initiation of a crack in low-cycle fatigue (LCF), adequate high-cycle fatigue (HCF) strength in resonance, creep rupture (CR) at elevated temperatures, or simply metal loss or recession due to oxidation. Typical applications envisaged for the PBLFT materials, at least initially, are those where the alloys are used in cast or wrought forms but not in applications that are safety-critical. For applications where strength is the main design consideration, IN-718, 15-5PH, and Ti-6-4 are viable candidates. For higher temperature applications, Co-Cr, IN-625, Hast-X, and IN-718 are the likely choices.

Table 3.1 A partial list of PBLFT alloys investigated at Honeywell Aerospace

Name	Final heat treatment parameters
IN-718	1352 K/1 h + 991 K/8 h + 894 K/8 h
IN-625	1255 K/1 h
Hast-X	1422 K/1 h
Co-Cr	1464 K/1 h
Ti-6-4	1033 K/2 h
15-5PH	1339 K/3 min + 825 K/4 h
316L	1339 K/1 h
AlSi10Mg	802 K/5 h + 433 K/12 h

LCF is a phenomenon which occurs mostly at stress concentrations and at stress ranges greater than 75% of the yield strength. The LCF design cycle is typically representative of the loads and temperatures associated with the typical start-to-stop cycle in a turbine engine. The component may need to be certified with an extremely low probability of failure ($<10^{-9}$/cycle) in the cyclic range of 10^3–10^5 repeated loads.

HCF, on the other hand, occurs in response to vibratory loads at a fairly high cyclic frequency (10^2–10^4 Hz) and at low stress ranges. The design requirement would be absence of failure due to random or resonant vibratory loads during the component's certified life.

The phenomenon of creep occurs at relatively higher temperatures. Avoidance of change in shape or dimension, or rupture over the life of the component so as not to compromise its intended function is the goal.

Damage tolerance signifies the ability of the material to sustain failure due to propagation of a crack initiated by any of the above scenarios through an inspection period. Here crack growth and fracture toughness properties dictate proper usage.

Oxidation is an important consideration at temperatures well above 922 K especially in components such as turbine cases, combustor liners, and accessories such as fuel injectors.

In addition to the information on microstructure, mechanical property data and fractographic information will be presented below. Where available, the data will be compared to published averages of the mechanical properties expected from wrought or cast versions.

3.4 Grain Size

All alloys in Table 3.1 except IN-625, Hast-X, Co-Cr, and 316L are precipitation strengthened. The rest derive their strength from solid solution hardening. In both cases grain size can have a significant influence on mechanical strength. Super alloys such as IN-718 are the most influenced by grain size as they have a planar mode of deformation in which slip is distributed heterogeneously within a grain and concentrated on fewer slip planes.

In processing of wrought products, recrystallization may take place during hot working or during subsequent solution heat treatment. The resultant grain size is determined by the level of prior hot or cold work, the level of thermal energy available for grain growth and the resistance to grain growth offered by grain boundary pinning particles. A dispersion of fine incoherent particles can limit the grain growth by the phenomenon called Zener pinning [1]. Since almost all of the materials shown in Table 3.1 have carbon as a constituent, carbide particles can appear as a fine dispersion in AM products. In addition, any remnant oxygen in the build atmosphere may create a dispersion of fine oxides in the matrix under certain conditions. If nitrogen is used as a build atmosphere it can contribute to further inhibition of grain growth by formation of fine nitride particles. Although in materials like IN-718 the HIP and solution temperatures are well above the solvus temperature of the strengthening precipitates, these fine carbide, nitride, or oxide particles present even at the super solvus temperature of the precipitates can limit grain growth to ASTM 3–5. In IN-625 and Co-Cr, vendor to vendor differences have been noticed in grain sizes brought about by differences in distribution of these grain boundary pinning inclusions.

3.5   Tensile Properties

Tensile properties of all alloys in the table were determined at selected temperatures. In general, they all compared well with the corresponding wrought material and were better than their cast versions. IN-718 was the most characterized of all as it is one of the most used alloys in gas turbine engine applications. Tensile, LCF, HCF, CR, fracture toughness, and cyclic crack growth data were generated for IN-718 at a variety of temperatures. A typical micrograph of PBLFT IN-718 is shown in Fig. 3.1 which had a grain size of ASTM 4. The particles which were identified as carbides of niobium or titanium are seen as a fine dispersion in the background and the distribution is fairly uniform across. Grain size and shape were found to be uniform in both transverse and longitudinal metallographic sections of the test bars. There were no discernible differences in microstructure between the bars built in vertical, horizontal or in 45° orientations, which attests to the capability of PBLFT process to impart homogeneous microstructure. Measurement of elastic modulii, which is a good indication of texture, showed no significant differences

Fig. 3.1 Microstructure of PBLFT IN-718

between horizontally built, vertically built or those built in 45° orientations. At 589 K they were respectively 185, 173 and 194 GPa, whereas for a fine grained wrought IN-718 [2] the reported value is 184 GPa. These results mean insignificant crystallographic texture in PBLFT IN-718 given even that the grain size was coarse compared to typical wrought IN-718 (ASTM 8–10) [2].

Figure 3.2 shows tensile strengths of PBLFT IN-718 for three orientations with respect to the build plate. Horizontal bars were tested at 297, 589, 700, 811, and 922 K, whereas vertical and 45° orientations were tested only at 700 and 922 K. The data are comparable to expected strengths for a grain size of ASTM 4. There were a total of seventy-two tensile bars tested. YS and UTS refer to 0.2% yield and ultimate tensile strengths, respectively. The variations between different orientations are not discernible on the average, which point to uniformity of texture across different build directions. The tensile elongation values were found to be slightly higher for the PBLFT product than for fine grained wrought product [2]. When compared to cast IN-718, tensile strengths and ductility of PBLFT IN-718 were far superior. Statistical analysis of the YS and UTS at 700 and 922 K (24 values for each temperature) which contained all three orientations showed a standard error of estimate of 33 MPa for YS and 27 MPa for UTS at *either* temperature. The fact that these values are somewhat lower than those observed in typical wrought IN-718, and much lower than those in cast versions speaks well of the uniformity that can be expected in a PBLFT process.

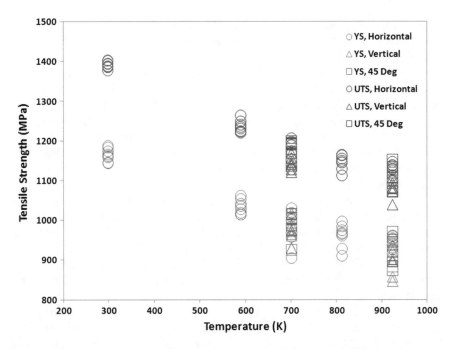

Fig. 3.2 Comparison of Tensile Strengths of PBLFT IN-718 for various build orientations

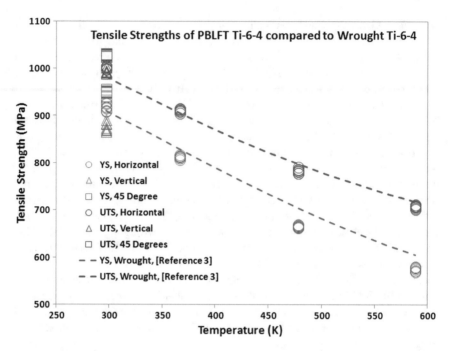

Fig. 3.3 Comparison of tensile strengths of PBLFT Ti-6-4 to wrought Ti-6-4

For Ti-6-4 alloy, the tensile strengths of PBLFT product compared well with the wrought version as seen in Fig. 3.3. The average lines for YS and UTS are taken from CINDAS [3]. This closeness to wrought behavior may be due first to the small grain size associated with PBLFT processing and subsequently to the fine and uniform lamellar structure (alpha-beta) observed in Fig. 3.4. The ductilities were higher than those for the wrought. Measurements of elastic modulii in horizontal, vertical, and 45° orientations showed no significant texture. For Ti-6-4 also, the tensile strengths and ductility were superior to the cast versions. The scatter in YS and UTS values, 31 and 15 MPa, at 297 K for 24 specimens equally distributed among the three orientations were similar to or lower than those observed for typical wrought products.

The PBLFT process is amenable to duplicating wrought properties in PH (precipitation hardening) steels like 15-5PH. Figure 3.5 shows a comparison. There are a total of 48 specimens with equal distribution with respect to the four temperatures and each temperature set containing equal distributions of the three orientations. The different orientations are not distinguished in the plot. The average properties in original brochures for the wrought version are shown in the figure by dashed lines [4].

For solid solution alloys like Hast-X, IN-625, and Co-Cr tensile properties approached those of the standard wrought product and in some cases even exceeded reported values. This is a direct result of inhibition of grain growth by dispersoids

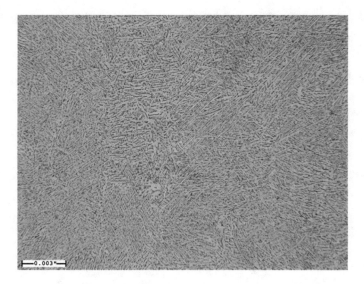

Fig. 3.4 Microstructure of PBLFT Ti-6-4

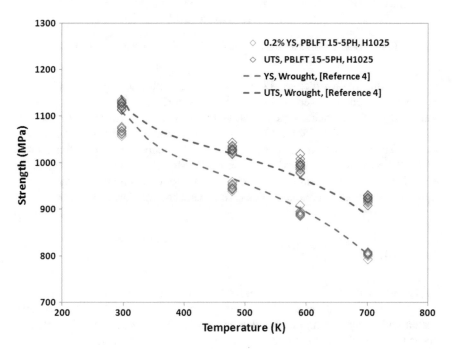

Fig. 3.5 Comparison of tensile strengths of PBLFT 15-5PH to wrought 15-5PH

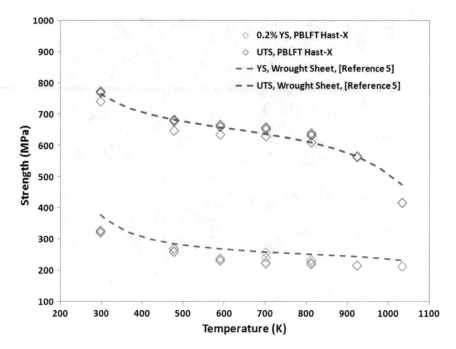

Fig. 3.6 Comparison of tensile strengths of PBLFT Hast-X to wrought Hast-X sheet

present in the microstructure. Figure 3.6 shows a plot of tensile properties of Hast-X compared to the reported wrought values in Haynes' brochure [5]. The data contained the three orientations at each temperature. Similar results were observed for IN-625 and Co-Cr. For Co-Cr the comparison was made against properties of wrought Haynes 188 which is a cobalt-based alloy widely used in high temperature applications in gas turbine engines. For AlSi10Mg alloy, tensile strengths of the PBLFT version were found to be superior to those for cast aluminum alloy 356, and equivalent to wrought aluminum 6061-T6 at temperatures lower than 366 K.

3.6 Fatigue Properties

For fatigue testing of all alloys except for IN-718, the specimen gage was machined to remove the as-built surface. In the case of IN-718, major part of the characterization was conducted on machined gage surfaces but a selected number of specimens were tested with the as-built surface in the gage section to ascertain the debit owing to defects present in the as-built layer.

Figure 3.7 shows a LCF plot of PBLFT IN-718 where the stress range on the ordinate is normalized to a stress ratio, R (min/max), of −1.0 by means of stress

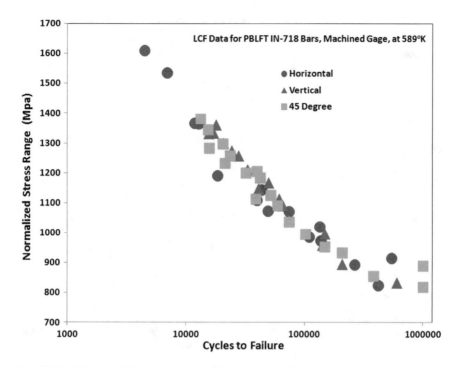

Fig. 3.7 LCF Data at 589 K for PBLFT IN-718 with machined gage section

ratio normalization factor similar to the Smith–Watson–Topper parameter [6]. The data covers strain ratios of 0.0 and −1.0 and three orientations of the bars with respect to the build plate. The LCF strengths seem to be independent of build orientations. Also a single stress ratio correction was found to be applicable for all three build orientations. This again speaks well of the uniformity of microstructure and lack of texture in all three build orientations. When the data are compared to those reported in the literature [7] for finer grained higher strength wrought IN-718 (ASTM 8–10) the difference is explainable simply by the grain size effect. Figure 3.8 shows the debit in LCF strength when the as-built surface is not removed from the gage. It is interesting to note that the effect is not evident in completely reversed cycles ($R = -1.0$). At more positive strain ratios the debit is well recognizable. Note that all data shown in Fig. 3.8 are from vertically built bars.

Figures 3.9 and 3.10 show respectively scanning electron micrographs of typical crack initiation in LCF for as-machined and as-built surfaces. Note the subsurface crystallographic Stage-I initiation in the machined surface and a more frontal, surface initiated, noncrystallographic Stage-II type for the as-built surface. The morphology in Fig. 3.9 is typical when the crack initiation is not associated with or influenced by defects. In Fig. 3.10 *the solid brush line* drawn near the bottom of the

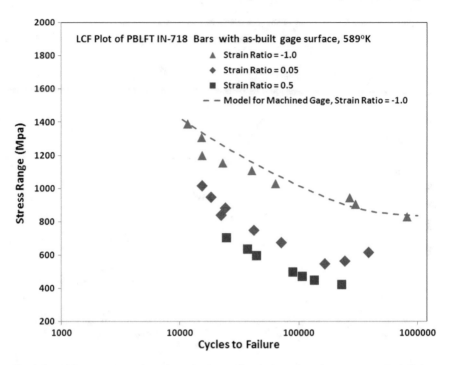

Fig. 3.8 LCF Data at 589 K for PBLFT IN-718 with as-built gage surface, all vertically built bars

fracture (*not to be mistaken for a crack*) demarcates the shape and location of the defect front in the plane of fracture from which fatigue initiated. This debit due to defects in as-built surface was also seen in HCF data generated at 589 K. HCF strength for specimens with machined gage surfaces was found to be equivalent to that for wrought version. This may be related to smaller slip distances involved in HCF crack initiation which may be influenced by dispersoids. An important point to note in both LCF and HCF behavior of PBLFT IN-718 is that even the strength with the as-built surface present in the gage was far superior to those of the cast versions. With new methods ever evolving in precision surface machining techniques, the as-built surface debit in LCF in PBLFT products may become a thing of the past.

The fatigue strength of Ti-6-4 was found to be equivalent to that for wrought and much superior to cast version. The former is attributed to the fine nature of the microstructure and the latter to the very large grain sizes of the cast. HCF strength of PBLFT Ti-6-4 was higher than that for wrought especially at higher mean stresses. For 15-5 PH and 316L, the LCF strength at 297 K fell between average and minimum (defined by −3Sigma) of wrought models. For PBLFT IN-625 and

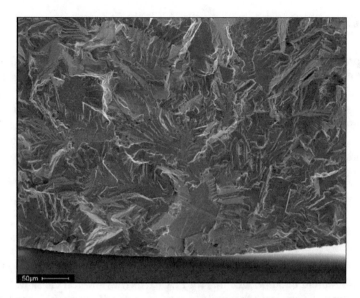

Fig. 3.9 SEM micrograph of crack initiation with machined gage section, vertically built, Nf = 206,544

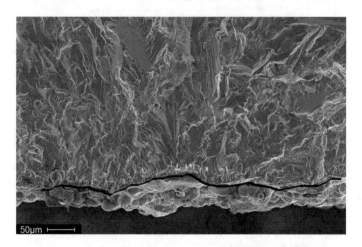

Fig. 3.10 SEM micrograph of crack initiation with as-built gage section, vertically built, Nf = 21,855. Note the rough as-built surface at the bottom edge and the solid brush line indicating the LCF initiation front

Hast-X, the fatigue strengths were found to be equivalent to those for wrought. For Co-Cr the LCF strength was far superior to that for wrought Hast-X, HA-230, or the cobalt base wrought alloy HA-188.

3.7 Creep

Creep properties were measured for PBLFT IN-718. For both rupture and times to various percentages of creep strains PBLFT version was superior to wrought or cast versions. The superiority to cast version was unexpected. Figure 3.11 shows a comparison of rupture times. PBLFT version significantly exceeded in capability over the cast version at 977 and 1033 K. This superiority in creep and rupture strengths of PBLFT IN-718 also may be attributed to the effective pinning of grain boundaries by dispersions. There were no orientation related differences in properties between different build directions.

3.8 Fracture Tolerance

Fracture toughness measured at room temperature for various orientations of the crack for different build orientations for PBLFT IN-718 showed the same values as those reported for wrought fine grained version in the literature. The crack growth rates measured for PBLFT IN-718 at 589 K were lower than those fine grained wrought version. This difference may be attributed to larger grain size of PBLFT product.

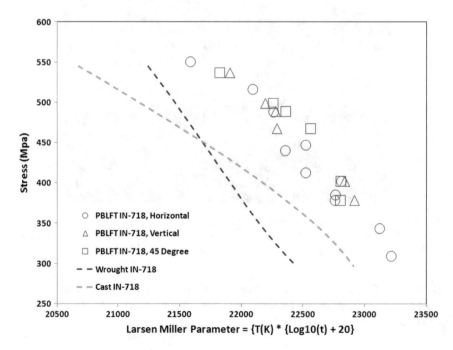

Fig. 3.11 Larson–Miller plots of CR for PBLFT IN-718

3.9 Influence of Dispersoids

A uniform distribution of fine incoherent particles in an alloy matrix will have an influence on keeping grain size from becoming too large in spite of heat treatments conducted at a high enough temperature to put into solution all other strengthening phases. This is somewhat akin to the behavior observed in super solvus processed PM (powder metallurgy) nickel base superalloys like LC PM Astroloy, Alloy-10, PM Rene95, and others. The grain growth inhibiting particles in the PM alloys are fine dispersion of carbides or oxides created during powder manufacture. In the case of AM processed IN-718 these particles are primarily carbides of niobium or titanium.

Vendor to vendor differences have been noted in tensile yield strength for IN-625, Co-Cr, and stainless steel 316L. Figure 3.12 shows the observation on PBLFT IN-625 in which the data are compared to standard wrought IN-625 [8]. The effect is explainable based on grain size differences and influence of dispersoids. In the case of Co-Cr, our investigation has shown that introduction of nitrogen (versus argon) into the build atmosphere can produce the same effect, with nitrogen build atmosphere contributing higher yield strength and higher creep strength. In addition, variations in processing parameters or the initial chemistry of the powder itself could affect grain growth.

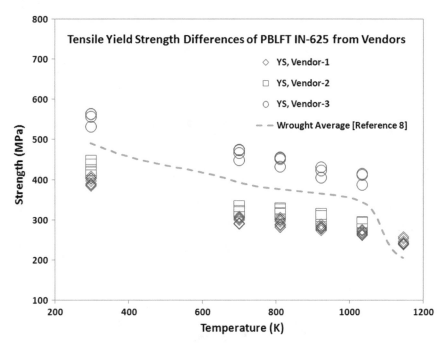

Fig. 3.12 Vendor differences in tensile yield strength of PBLFT IN-625

3.10   Electron Beam Technology

AM using electron beam technology (EBT) has begun to get attention for production of aerospace components. IN-718 has been used as a vehicle at Honeywell for understanding the unique microstructure and properties produced by EBT. Because of the higher thermal intensity of the beam and that the base plate has to be held at a high enough temperature (1200–1350 K) during manufacture, the vertical growth of the successively deposited layers seems to occur highly epitaxially. As a result, there is a strong and significant texture developed in the vertical (build) direction. LCF tests have shown that the modulus is close to that for directionally solidified nickel base superalloys like DS MAR-M-247 in the growth direction, which is also close to the modulus value for nickel base superalloy single crystals in the $\langle 001 \rangle$ direction. Figure 3.13 shows the data for heat treated EBT IN-718. Modulus of PBLFT IN-718, however, is close to that for wrought IN-718.

Microstructure of EBT IN-718 consists of grains with a high aspect ratio in the build direction as shown in Fig. 3.14. This texture was seen to exist even in the as-built condition. Electron back scattered diffraction (EBSD) orientation map of these grains is shown in Fig. 3.15. The strong red coloring [9] is indicative of the

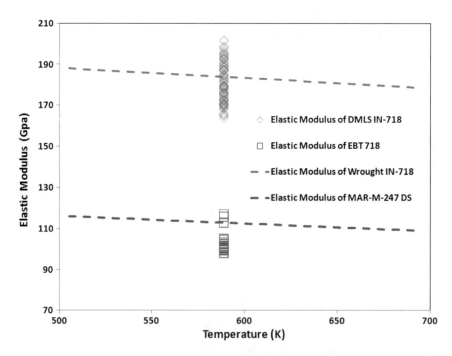

Fig. 3.13 Comparison of elastic modulii of EBT IN-718 and PBLFT IN-718

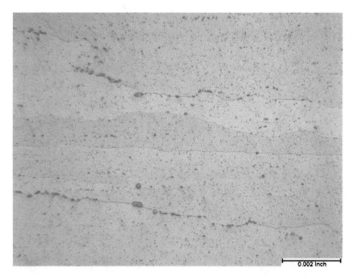

Fig. 3.14 Longitudinal section (*Horizontal direction in the Micrograph, also the Build Direction*) of heat treated EBM IN-718 showing grains with a high aspect ratio

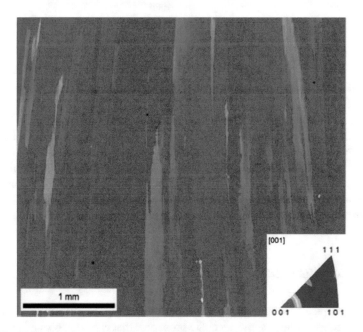

Fig. 3.15 EBSD crystallographic Orientation Map of EBT IN-718 (*The vertical orientation here is the build direction, and corresponds with the horizontal direction in Fig. 3.14*)

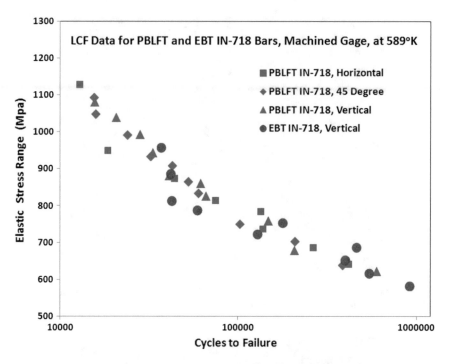

Fig. 3.16 LCF data of EBT IN-718 compared to PBLFT IN-718 at 589 K, strain ratio of 0.0

preferred [100] crystallographic direction in the build orientation (vertical in the figure).

In spite of the strong texture, the tensile and LCF strengths of EBT IN-718 were close to those for the PBLFT material. Figure 3.16 shows a comparison of the LCF data. The EBT data fall on top of the PBLFT version. Because of the low elastic modulus in the build direction, EBT process may be suitable for situations where thermally induced stresses are predominant in the build direction of a component. The fractography of EBT bars resembled those observed for DS MAR-M-247 with significant Stage I crystallographic faceted crack initiation.

3.11 Conclusions

The general picture that emerges from the initial material evaluation is that AM produced metallic alloys show adequate properties that can compete and potentially replace their cast version equivalents, even with their as-built surface. The inherent AM material fatigue capability, when the as-built surface is removed, has been found to be almost commensurate with the wrought versions for all solid solutioned alloys and for Ti-alloys. Continued developments and improvements in AM processing will likely provide additional flexibility in tailoring and optimizing material microstructures and the resultant mechanical properties.

References

1. Hillert M (1988) Inhibition of grain growth by second-phase particles. Acta Metall 36 (12):3177–3181
2. Inconel Alloy 718, Special Metals Brochure. Huntington, 3200 Riverside Drive, WV 25705-1771
3. Aerospace and High Performance Alloys Database, CINDAS, Data digitized from Figure 3.3.1.8, Ti-6-4 Chapter, Section 3707, November 2000
4. Product Bulletin No. S-6, Armco 15-5 PH VAC CE Precipitation Hardening Steel, Armco Stainless Steel Products, Baltimore, MD 21203
5. Haynes Brochure, Hastelloy X Alloy, Haynes International, 1020 West Park Avenue, Kokomo, IN 46904
6. Smith KN, Watson P, Topper TH (1970) A stress-strain function for the fatigue of metals. J Mater ASTM 5:767–778
7. Korth GE (1983) Mechanical properties test data of alloy 718 for liquid metal fast breeder reactor applications, Report EGG-2229, Idaho National Engineering Laboratory, January 1983
8. Haynes Brochure, IN-625 Alloy, Haynes International, 1020 West Park Avenue, Kokomo, IN 46904
9. EBSD Work was Conducted at Arizona State University under the guidance of Professor Amaneh Tasooji and the students of Capstone Project, May 2016

Chapter 4
Electron Beam Technology

4.1 Additive Manufacturing

Additive manufacturing (AM) or three-dimensional (3D) printing is the process of creating 3D objects or products, layer by layer, from a 3D digital model. The 3D digital model can be created from CAD, CT, MRI, or laser scanning. AM is the reverse process of traditional manufacturing technologies, such as machining, in which one starts with a block of material and subsequently, the material is removed in a subtractive process forming the final desired part. Alternatively, AM starts with a blank platform and material is added in a controlled method where required—in each layer—until the final part is formed [1].

In order to help standardize AM in the United States, the ASTM F42 Committee on AM Technologies was formed in 2009 and categorized AM technologies into seven categories including vat photopolymerization, material extrusion, powder bed fusion, material jetting, binder jetting, sheet lamination, and directed energy deposition (F42 Committee 2012). In order to provide sufficient background for the research completed in this dissertation, this chapter provides more information on the metal AM processes that are commonly used today in AM.

4.1.1 Powder Bed Fusion

Powder bed fusion systems use thermal energy to melt the powder into the desired pattern. The majority of powder bed fusion systems use a laser to melt polymer or metal powder to fabricate the 3D structures layer by layer (F42 Committee 2012). Powder bed fusion technology is the most common metal-based technology used to manufacture end use engineered products, many of which are being used in aerospace, defense, and medical applications.

© Springer International Publishing AG 2017
L. Yang et al., *Additive Manufacturing of Metals: The Technology, Materials,*
Design and Production, Springer Series in Advanced Manufacturing,
DOI 10.1007/978-3-319-55128-9_4

4.1.2 Electron Beam Melting

The electron beam melting (EBM) AM process builds complex parts out of metal powders within a high temperature and vacuum environment. EBM has been used in final production for many years, and more than 50,000 EBM-built devices have been implanted for the medical industry. In recent years, EBM has started to gain momentum in the aerospace industry to manufacture production components. In order to expand the EBM market, new materials such as Alloy 718 have been developed and characterized in the EBM technology. Alloy 718 is a

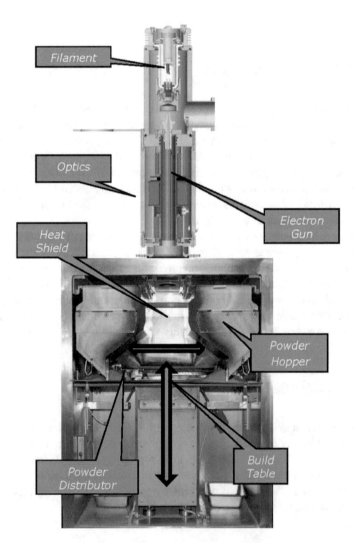

Fig. 4.1 EBM system

nickel-chromium-based super alloy ideal for high temperature and corrosive environments, with excellent mechanical properties at elevated temperatures.

Electron beam melting (EBM) is an AM technology that selectively consolidates metal powders such as titanium, Inco 718 and cobalt alloys to fabricate 3D structures. Compared to conventional manufacturing processes, the EBM process is capable of fabricating low-volume, high-value articles at reduced lead times. The technology was invented in 1993 in Sweden at the University of Technology in Gothenburg. Arcam was founded in 1997 and sold its first commercial system in 2002. The company currently has over 200 systems installed around the world [2, 3].

The process includes the focusing of an electron beam at discrete areas within a powder bed composed of metal particles (average size of 45–105 μm) to produce melting, followed by re-solidification that enables fabrication of complex geometries. The EBM system (Fig. 4.1) consists of an electron beam gun, vacuum chamber (∼10–4 torr), build tank, and powder distribution mechanisms (powder hoppers and rake). Within the electron beam gun, a tungsten filament is heated to emit electrons accelerated at high voltage (60 kV), resulting in an electron beam (carrying high kinetic energy) that is focused by electromagnetic lenses. This system is capable of electron beam scan speeds of up to 8000 m/s, electron beam positioning accuracy of ±0.025 mm, and layer thicknesses in the range 0.05–0.2 mm [4–6].

Part fabrication is initiated by the uniform distribution of powder over a start plate as seen in Fig. 4.2a. The powder layer is preheated by the electron beam followed by a sequence of line-by-line scanning that melts the loosely joined powder. The powder is spread and packed, only by the raking system; therefore, the powder is disorganized with gaps. As a result, before melting, the layer thickness is

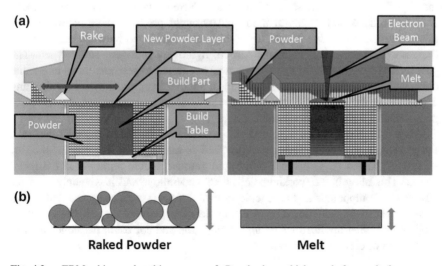

Fig. 4.2 a EBM raking and melting process. **b** Powder layer thickness before and after

about 2–3 times greater than it should be. When the powder is melted, it reduces to the correct layer thickness (Fig. 4.2b). After completion of a layer, the process platform is lowered a distance equivalent to one layer thickness (0.05–0.2 mm). A new layer of powder is applied and the process is repeated until the full build is complete. Completion of the process is followed by a helium-assisted cool-down sequence lasting about 6 h, depending on the build size and material used.

4.1.3   Materials

Titanium and associated alloys have been used in industry for over 50 years— becoming more popular in recent decades [7]. Titanium has been most successful in areas where the high strength to weight ratio provides an advantage over aluminum and steels. Other advantages of titanium include the ability to be biocompatible and corrosion resistant [8]. The new lighter and more efficient airplane designs require stronger frames and thus the obvious replacement for aluminum structural components is titanium; since aluminum and composites cannot provide the required structural support that titanium can provide. As composites become more popular in the aerospace industry, more designers employ titanium airframe components [9] Commercial (99.2% pure) grades of titanium have ultimate tensile strength of about 63,000 psi (434 MPa), equal to that of common, low-grade steel alloys, but with the dramatic advantage of being 45% lighter. Titanium is 60% more dense than aluminum, but more than twice as strong as the most commonly used 6061-T6 aluminum alloy [10].

Titanium is the ninth most abundant element in the Earth's crust (0.63% by mass) and the seventh most abundant metal. Current titanium production methods used to produce both titanium ingot and wrought products have naturally high manufacturing costs, due in part to the high chemical bond between titanium and the associated interstitial elements carbon, hydrogen, nitrogen, and oxygen. The cost increases significantly due to the energy required to separate these interstitial elements [11]. Ti-6AL-4V (Ti64) is the most popular titanium alloy for medical and aerospace applications [12] and the costs for titanium have increased due to recent high demand from these industries.

Alloy 718 is a solution annealed and precipitation hardened material designed for high-temperature applications. Alloy 718 is hardened by the precipitation of secondary phases (e.g., gamma prime and gamma double-prime) into the metal matrix. This alloy is the workhorse of nickel superalloys and it is widely used for gas turbine components and cryogenic storage tanks, jet engines, pump bodies and parts, rocket motors and thrust reversers, nuclear fuel element spacers, hot extrusion tooling, and other applications requiring oxidation and corrosion resistance as well as strength at elevated temperatures.

4.1.4   Powder Metallurgy Requirements for EBM

Before attempting to develop parameters using an EBM system, powder must be characterized to determine if it will be a good candidate for the technology. The powder having flowability (it must be able to flow 25 s/50 g like Arcam's supplied powder), high apparent density (>50% of the density of solid material), no internal porosity (from the production process), and contain no small particles (<0.010 mm) is preferred.

In order to determine the flowability of a powder, B212-09 ASTM International Standards were followed. The powder was allowed to flow freely and unaided through a Hall flowmeter funnel from ACuPowder International (Union, NJ) (shown in Fig. 4.3a), consisting of a calibrated orifice and a cylindrical brass cup with a nominal capacity of 25 cm^3 and a distance of ~25 mm from the bottom of the funnel to the top of the density cup [13]. Apparent density, flow rate, and percent density change were obtained for different powders and compared to Ti64 powder provided by Arcam. High flowability is recommended to improve the hopper's feeding and to obtain a more stable raking—both providing smoother powder layers. If the percent density change of the material to be analyzed is comparable, or higher, then the value obtained for Ti64 the microstructure can be analyzed and the particle size distribution can be measured. Powder examination can be performed through the use of a stereomicroscope to identify the shape of the

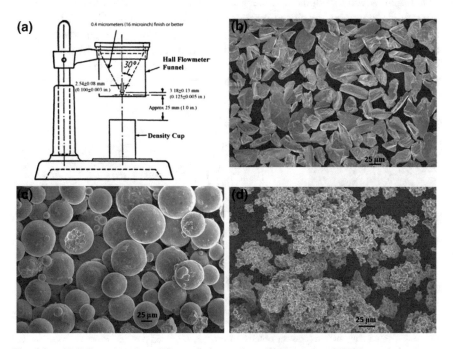

Fig. 4.3   a Hall flowmeter funnel. **b** tantalum powder. **c** Ti64 powder. **d** CPTi powder

particles, the presence of satellites, any foreign particles, or contamination. Figure 4.3b shows a scanning electron microscope (SEM) micrograph of tantalum powder with a 35% density change where angular morphology can be observed. Figure 4.3c depicts Arcam's Ti64 powder with a 55% density change and Fig. 4.3d illustrates a sponge-like commercially pure titanium (CPTi) powder with a 16% density change. Higher apparent densities will provide better heat conduction and therefore, reduce the risk of sample swelling or overheating. A higher value also improves the quality of supports. It can be observed that powders with undesirable morphologies (angular or irregular) have similar flowing capabilities [14].

Another aspect to consider when introducing a new powder to the EBM system is its internal porosity. Internal porosity is represented by spherical voids located along the inside of a single powder particle and can be analyzed by looking at a sample's cross section.

4.2   Powder Manufacturing

4.2.1   Gas Atomization

The process traditionally utilized to obtain spherical powder is gas atomization, which has existed since 1872 and was first patented by Marriot of Huddersfield. Since then, several designs have been used including different versions known as "free fall", "confined" or closed nozzles, and "internal mixing" [15]. This process, as depicted in Fig. 4.4, consists of dispersing liquid metal by a high velocity jet of air, nitrogen, argon, or helium. Many materials can be spheroidized by this process, such as copper and copper alloys, aluminum and its alloys, magnesium, zinc, titanium and its alloys, nickel-based alloys, cobalt-based alloys, lead/tin solder,

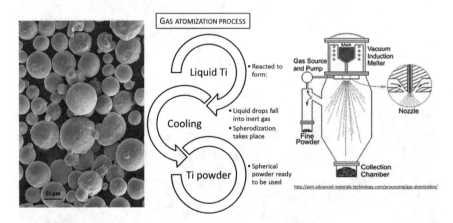

Fig. 4.4 SEM image of Ti64powder obtained from gas atomized powder. Process procedure and diagram inserted next to image

precious metals, etc. Inert gases must be used when atomizing reactive metals such as titanium and superalloys, to avoid oxidation [16]. Figure 4.4 shows an SEM image obtained from a Ti64 alloy that was gas atomized and purchased for EBM fabrication. In the case of titanium, the metal travels as a falling stream with the aid of gravity for 100–200 mm. Maximizing the gas's velocity and density when it meets the metal stream, allows one to achieve particle sizes in the range of 40–60 μm [17]. The schematic depicts a confined gas atomization system.

4.2.2 Induction Plasma Atomization

Induction plasma processes consist of in-flight heating and melting of feed material particles by plasma, followed by solidification under controlled conditions. Depending on the size and apparent density of the treated powder, the time of flight of the particle is controlled such that the molten droplets have sufficient time to complete solidification before reaching the bottom of the reactor chamber (www. tekna.com). The particle morphology and size distribution can be controlled by modifying the nozzle design, particle exiting velocity from the nozzle, and drop solidification distance [18].

The induction plasma process is used to improve powder characteristics such as increased flowability, decreased porosity, increased powder density, and enhanced powder purity. Flowability is increased when the particles are more spherical, making them easier to flow. Porosity is reduced when the material is remelted and resolidified. Powder tapped density increases since the re-spheroidized particles are more uniform, improving packing. The melting process can also be used to improve powder purity through the selective/reactive vaporization of impurities by increasing the plasma melting temperature [19]. Figure 4.5 shows re-spheroidized International Titanium Powder-Cristal Global (ITP) CPTi powder and the induction plasma process.

4.2.3 Armstrong Process

The Armstrong process provides high purity metal and alloy powders at a lower cost and is particularly well suited for titanium. The main aspect of this process is the elimination of the traditional Kroll process which is responsible for producing the unintentional "sponge" morphology. Alternatively, the Armstrong process starts producing powder to be consolidated subsequently, and therefore, powder manufacturing cost is reduced through the minimization of processing steps [20]. The Armstrong process is owned by ITP and can be used for a wide range of metals, alloys, and ceramics (www.itponline.com). Among the many benefits of this process are high purity, controllable oxygen content, uniform grain structure, and cost reduction [21, 22]. More importantly, ITP powder has been shown to meet ASTM

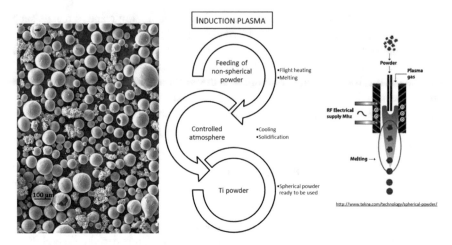

Fig. 4.5 SEM image of Armstrong Ti powder after induction plasma process. Diagram and schematic of the process illustrated next to image

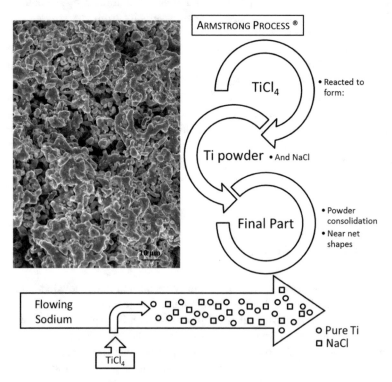

Fig. 4.6 SEM image of pure Ti sponge powder obtained from the Armstrong process. Diagrams illustrating the process are shown adjacent to image

grade 2 and ASTM grade 5 properties for CPTi and Ti64 [23] Ti and Ti64 powders obtained from the Armstrong process have been utilized in the production of roll compacted sheets and vacuum hot pressed plates with appropriate properties [24]. Figure 4.6 is an SEM image of the sponge-like titanium obtained via the Armstrong process as well as the sequence of steps necessary to obtain such powder. Titanium tetrachloride (TiCl4) is reacted with sodium to form pure titanium and sodium chloride (NaCl).

Titanium powder from the Armstrong process is comparable to CPTi ASTM Grade 2 with regards to oxygen and carbon content since both are within the ASTM range. In Armstrong research, Armstrong Ti powder was compacted by a specialized vacuum hot press (VHP) unit and tensile testing was performed. UTS and YS are higher than typical CPTi Grade 2 due to the all-alpha, fine, and equiaxed grain structure obtained, $\sim$2–4 Å [25].

4.2.4 Hydride-Dehydride

The hydride–dehydride (HDH) process has the capability to produce titanium, zirconium, vanadium, and tantalum powders. The process, as seen in Fig. 4.7, consists of performing a reversible reaction on the material by hydrating titanium at

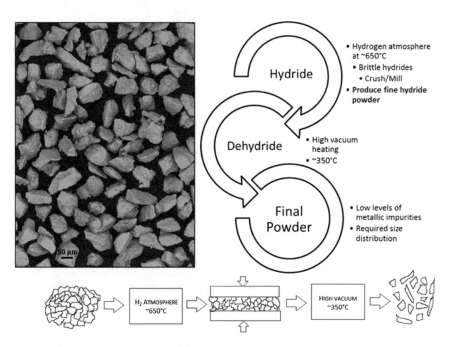

Fig. 4.7 SEM image of pure Ti64 powder obtained from the HDH process. Diagrams illustrating the HDH process are shown adjacent to image

elevated temperatures ($\sim$650 °C) for long periods of time, allowing the brittle phase of the titanium to be crushed and milled for finer powder size. The crushed powder is then reheated under high vacuum at $\sim$350 °C to remove the hydrogen [26]. The result is an angular shaped (nonspherical) powder that can be used successfully for fabrication in traditional powder production. The traditional methods for utilizing HDH powder are press sintering, metal injection molding, cold isostatic pressing, and hot isostatic pressing [27].

4.3 Powder Characterization

Prior to parameter development using the EBM system, the powder must be characterized to determine if it is a good candidate for the technology. The steps in Fig. 4.8 were followed for initial powder analysis. In order to determine if the powder was safe to handle and use in the EBM system, a minimum ignition energy test was performed following standard BS EN 13821:2002 (Determination of minimum ignition energy of dust/air mixtures) [26]. The test measured the ease of ignition of a dust cloud by electrical and electrostatic discharge. Electrodes were connected to a circuit that produced a spark with a known energy. A cloud of powder was blown past the spark; if the powder ignited, the energy was reduced until there was no reaction. Arcam recommends powder with 45–105 µm particle distribution that have minimum ignition energy of 0.5 J.

The powder must have high flowability (it must be able to flow 25 s/50 g, like Arcam's supplied powder) to be able to rake correctly and flow in the machine hopers. Following ASTM standard B213, a Hall flow meter was used to test

Fig. 4.8 Process flow diagrams showing steps for initial powder analysis

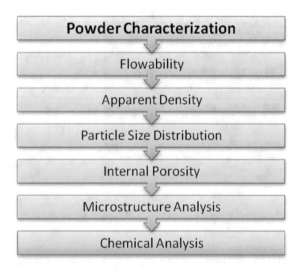

flowability [13]. The powder must also have high apparent density (apparent density must be >50% of the density of solid material) to be able to melt the powder and obtain fully dense parts with current scanning methods. Following standard ASTM B212, a Hall flow meter and a density cup were used to test apparent density [13]. The particle size distribution was verified as Arcam recommends that particle distribution be between 45 and 105 μm with a normal distribution curve. The powder must not contain small particles (<0.010 mm) since these can become a fi and health hazard. Next, the powder was mounted and polished to check for internal porosity (from the gas atomization production process). Internal porosity can become problematic with various materials, like Ti64, that do not allow entrapped gas to flow out of the part as the powder melts and solidifies.

Microstructure analysis of the powder was performed to obtain a basis of comparison before and after the powder was processed with EBM technology. Chemical analysis was also performed to compare the powder before and after melting by the EBM process, since the electron gun's energy can evaporate a percentage of the lower melting element in the material. This can put the material out of specification, making the end part non-compliant. These powder analyses can save a lot of time in determining if the powder will be usable in EBM systems, stopping one from using a powder that has the incorrect packing or flowability density.

An Alloy 718 powder selection study has been done with ARCAM and ORNL. The study involved powders from nine different powder manufacturers using four different manufacturing processes. The morphology of the gas atomized powders to the rotary atomized, plasma atomized, and the plasma rotated electrode.

4.4 Parameter Development

In order to start the material parameter process development (Fig. 4.9), a smoke test was performed in the build chamber with cold powder. Smoke takes place when the charge distribution density exceeds the critical limit of the powder. As a result, an electrical discharge occurs since the powder particles repel each other, causing a powder explosion inside the chamber, as seen in Fig. 4.10. The smoke test consisted of using various machine parameter settings of line order, line offset, beam speed, and focus offset to determine which settings were within the "smoke-free" parameter window. Once the parameter window was identified, the parameters could be modified safely. If the parameters are within the window, the build should be smoke free.

Next, a sintering test was conducted to identify the temperature required to preheat the build platform to sinter powder underneath the platform. The sintering test consisted of preheating a clean build platform inside the machine to a constant

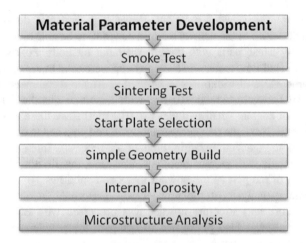

Fig. 4.9 Process flow diagrams showing steps for initial process development

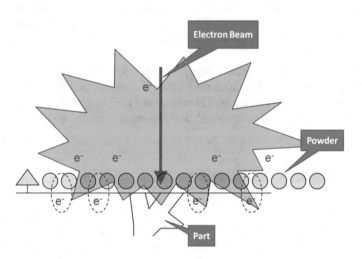

Fig. 4.10 The charge distribution density exceeds the critical limit of the powder, due to the electron beam, and the powder repels each other creating an explosion inside the build chamber

temperature and holding it for 20 min. The build was stopped and the chamber was opened to check if the powder underneath the build platform was sintered. If the powder was not sintered, the test was repeated at a higher temperature. The powder must be sintered for platform stability when raking as well as for thermal conductivity. The advantage of the sintering test is that it provides an indication of the current required for preheating to maintain a stable temperature during the build. The start plate material must be compatible with the new powder material since the start plate holds the powder in place during a build. Stainless steel has been the most successful start plate material as it is compatible with a wide selection of build materials.

Once the correct start temperature and start plate materials were identified, the first process build was prepared. The first build must be a simple geometry, like blocks with no negative surfaces or supports. Initially, the Ti64 themes provided by Arcam can be used as a good starting point. There are three main build themes for every build: the preheat theme, the melt theme, and the support theme. For every build step, there is a theme that provides the parameters to the machine. The initial themes were modified to reflect the findings in the smoke test, so that they were within the parameter window, to eliminate smoke during the build. Next, the preheating theme was modified by taking the average current obtained in the sintering test and using it as the beam current. The advantage of the Arcam system is that theme parameters can be modified while the system is building.

To begin construction, the Arcam system must be prepared. Next, the build start plate was insulated by placing powder underneath and leveled. After vacuuming down the chamber, the build plate was preheated to the temperature determined by the sintering test. In order to understand and control how the system builds, the process was viewed through the chamber window. Parameters that produced a smooth melting, without fireworks, were found. Fireworks occur when the powder explodes as the electron beam is melting it. Fireworks can be eliminated or reduced by using a higher process temperature or increasing the current in the preheat process.

Powder sintering was controlled by the preheating theme parameters: min/max current, focus offset, and number of preheating repetitions. The preheating current controls how much energy is delivered during one cycle run of the heating process. The preheat focus offset controls the electron beam diameter during preheating. The focus offset parameter provides the area where energy is being distributed and can be set to deliver maximum energy without hard-sintering the material. The number of repetitions conveys how many times the build is preheated and is used to raise and stabilize the bed temperature. These parameters control the temperature and stabilize the powder before the melting process begins.

The melt process is when the powder is liquefied layer by layer forming the final part. During this process, the powder surface must look even and smooth to obtain fully dense parts. The melt process was controlled by the melt speed function and focus offset. The speed function value controlled the melting speed in automatic mode. Higher values provide higher speed and deliver less energy when melting. Once again the focus offset was modified to control the melt providing an even and smooth top surface. Focus offset calibration was done by modifying the focus offset in each cube to obtain the best top surface finish, as seen in previously built samples in Fig. 4.11, where the best surface is the one circled in red. The value selected was used as the focus value for the rest of the builds.

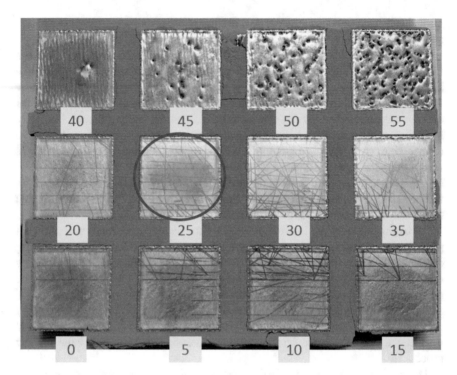

Fig. 4.11 Focus offset test for selecting best focus offset (*circled in red*)

4.5 Build Setup and Process

Once the parameters were determined to be sufficient, a verification build was manufactured using a 150 × 150 mm 316 stainless steel start plate. The verification build (Figs. 4.12 and 4.13) consisted of four rectangular blocks of 100 × 25 mm

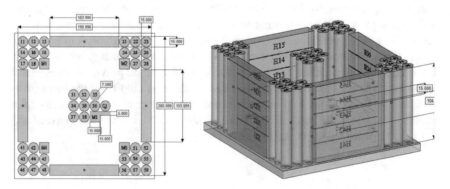

Fig. 4.12 Arcam verification build 1

Fig. 4.13 Verification build 1 made in a 150 × 150 mm start plate

and 32 cylinders of 10 mm in diameter and 105 mm in height. Additionally, four cubes are also included for chemistry and microstructure analysis. The stainless steel plate was given an initial plate preheat temperature of 1000 °C and a sinter temperature of ≈20 min. Overall, the build time was ≈30 h with a ≈6 h period required for cool-down of the build. For the present build, a layer thickness of 75 μm was chosen as a compromise between build speed and surface quality.

References

1. Gibson I, Rosen DW, Stucker B (2010) Additive manufacturing technologies: rapid prototyping to direct digital manufacturing. Springer, Berlin
2. www.arcam.com
3. Wohlers T (2012). Wohlers Report 2012. May. https://lirias.kuleuven.be/handle/123456789/372635
4. Gaytan SM, Murr LE, Medina F, Martinez E, Lopez MI, Wicker RB (2009) Advanced metal powder based manufacturing of complex components by electron beam melting. Mater Technol: Adv Perform Mater 24(3):180–190. doi:10.1179/106678509X12475882446133
5. Murr LE, Gaytan SM, Medina F, Lopez H, Martinez E, Machado BI, Hernandez DH et al (2010) Next-generation biomedical implants using additive manufacturing of complex, cellular and functional mesh arrays. Philos Trans R Soc A: Math, Phys Eng Sci 368(1917): 1999–2032. doi:10.1098/rsta.2010.0010 April 28

6. Murr LE, Martinez E, Gaytan SM, Ramirez DA, Machado BI, Shindo PW, Martinez JL et al
    (2011) Microstructural architecture, microstructures, and mechanical properties for a
    nickel-base superalloy fabricated by electron beam melting. Metall Mater Trans A 42
    (11):3491–3508. doi:10.1007/s11661-011-0748-2
7. Boyer RR, Briggs RD (2005) The use of β titanium alloys in the aerospace industry. J Mater
    Eng Perform 14(6):681–685. doi:10.1361/105994905X75448
8. Azevedo CRF, Rodrigues D, Beneduce Neto F (2003) Ti–Al–V powder metallurgy (PM) via
    the hydrogenation–dehydrogenation (HDH) Process. J Alloys Compd 353(1–2):217–227.
    doi:10.1016/S0925-8388(02)01297-5 April 7
9. Peters M, Kumpfert J, Ward Ch, Leyens C (2003) Titanium alloys for aerospace applications.
    Adv Eng Mater 5(6):419–427. doi:10.1002/adem.200310095
10. Cui C, Hu BM, Zhao L, Liu S (2011) Titanium alloy production technology, market prospects
    and industry development. Mater Des 32(3):1684–1691. doi:10.1016/j.matdes.2010.09.011
11. Ivasishin OM, Anokhin VM, Demidik AN, Savvakin DG (2000) Cost-effective blended
    elemental powder metallurgy of titanium alloys for transportation application. Key Eng Mater
    188:55–62. doi:10.4028/www.scientific.net/KEM.188.55
12. Baufeld B, Brandl E, van der Biest O (2011) Wire based additive layer manufacturing:
    comparison of microstructure and mechanical properties of Ti–6Al–4V components
    fabricated by laser-beam deposition and shaped metal deposition. J Mater Process Technol
    211(6):1146–1158. doi:10.1016/j.jmatprotec.2011.01.018 June 1
13. Standard ASTM (2007) B212-99 'Standard test method for apparent density of free-flowing
    metal powders using the hall flowmeter funnel'. Annu Book ASTM Stand 2:23–25
14. Santomaso A, Lazzaro P, Canu P (2003) Powder flowability and density ratios: the impact of
    granules packing. Chem Eng Sci 58(13):2857–2874
15. Lee PW, Trudel Y, German RM, Ferguson BL, Eisen WB, Mover K, Madan D, Sanderow H,
    Lampman SR, Davidson GM (1998) Powder metal technologies and applications. Prepared
    under the direction of the ASM International Handbook Committee. ASM Handb 7:1146
16. Neikov OD, Naboychenko S, Mourachova IB, Gopienko VG, Frishberg IV, Lotsko DV
    (2009) Handbook of non-ferrous metal powders: technologies and applications. Elsevier,
    Amsterdam
17. Cui C, Hu BM, Zhao L, Liu S (2011) Titanium alloy production technology, market prospects
    and industry development. Mater Des 32(3):1684–1691. doi:10.1016/j.matdes.2010.09.011
18. Boulos M (2004) Plasma power can make better powders. Met Powder Rep 59(5):16–21.
    doi:10.1016/S0026-0657(04)00153-5
19. www.tekna.com
20. Stone N, Cantin D, Gibson M, Kearney T, Lathabai Sri, Ritchie David, Wilson R, Yousuff M,
    Rajakumar R, Rogers K (2009) Implementing the direct powder route for titanium mill
    product: continuous production of CP sheet. Mater Sci Forum 618–619:139–142. doi:10.
    4028/www.scientific.net/MSF.618-619.139
21. Chen W, Yamamoto Y, Peter WH, Gorti SB, Sabau AS, Clark MB, Nunn SD et al (2011)
    Cold compaction study of armstrong process® Ti–6Al–4V powders. Powder Technol 214
    (2):194–199. doi:10.1016/j.powtec.2011.08.007
22. Froes FH (1998) The production of low-cost titanium powders. JOM 50(9):41–43.
    doi:10.1007/s11837-998-0413-4
23. Peter WH, Blue CA, Scorey CR, Ernst W, McKernan JM, Kiggans JO, Rivard JDK, Yu C
    (2007) Non-melt processing of 'low-cost', armstrong titanium and titanium alloy powders.
    Proceedings of the 3rd international conference on light metals technology. Saint-Sauveur,
    Quebec, pp 24–26
24. Rivard JDK, Blue CA, Harper DC, Kiggans JO, Menchhofer PA, Mayotte JR, Jacobsen L,
    Kogut D (2005) The thermomechanical processing of titanium and Ti64 thin gage sheet and
    plate. JOM 57(11):58–61. doi:10.1007/s11837-005-0029-x

25. Eylon D, Ernst WA, Kramer DP (2013) Ultra-fine titanium microstructure development by rapid hot-compaction of armstrong-process powder for improved mechanical properties and superplasticity. http://www.itponline.com/docs/Eylon_Plansee_2009_TiPM_Paper.pdf. Accessed 30 April

26. McCracken CG, Motchenbacher C, Barbis DP (2013) Review of titanium. Powder-production methods. Int J Powder Metall 46(5):19–26 April 15

27. McCracken CG, Barbis DP, Deeter RC (2011) Key characteristics of hydride-dehydride titanium powder. Powder Metall 54(3):180–183. doi:10.1179/174329011X13045076771849

Chapter 5
Design for Additive Manufacturing

5.1 Overview

As AM technologies become increasingly recognized, manufacturing industries have started to consider AM as one of the enabling technologies that will transform the paradigm of manufacturing. In aerospace industries, companies have embraced AM in the belief that it will become critical to their competitiveness [1–3]. In biomedicine industries, AM is considered to be the most promising manufacturing technology to achieve customized medicine and treatment in the future [4, 5]. In addition, with digitalized information exchange and process management, AM is also regarded integral to the so-called "Industry 4.0" revolution, which enables intelligent and agile manufacturing via the integration with various tools including internet of things and machine–human connectivity and interactions [6, 7].

The successful adoption of AM requires not only extensive understanding of the process principles and machine characteristics, but also knowledge of how these processes can be effectively utilized to realize designs and functionality. It is now widely recognized that AM overcomes various limitations with geometry realization compared to traditional manufacturing methods such as machining, forging, casting and welding. Complex designs such as patient-specific hip implants and topology optimized bicycle frame that are difficult to realize via traditional manufacturing can be more readily fabricated via AM. On the other hand, it should also be realized that AM technologies are also subject to various manufacturing limitations. For example, even though in theory calibration structures with internal conformal cooling channels can be readily fabricated via powder bed fusion AM processes (Fig. 5.1), in reality this is often impractical due to the inaccessibility of the loose powder that is trapped in the cooling channels. Also, for parts with large overhanging areas, the need for support structures might compromise the manufacturability of the parts since the support structures often require labor-intensive manual removal processes (Fig. 5.2). In addition, similar to some traditional manufacturing processes such as casting and welding, many AM processes

© Springer International Publishing AG 2017
L. Yang et al., *Additive Manufacturing of Metals: The Technology, Materials, Design and Production*, Springer Series in Advanced Manufacturing, DOI 10.1007/978-3-319-55128-9_5

Fig. 5.1 Calibration tool
with conformal cooling
channel designs [8]

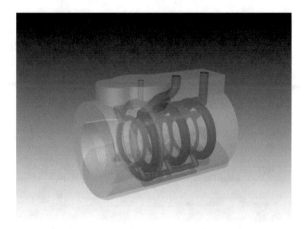

Fig. 5.2 Comparison
between CAD and actual part
with direct metal laser
sintering (DMLS) [9]

integrate the geometry generation and the functionality generation (e.g. mechanical properties, physical properties) in one single manufacturing step. Therefore, during the design of the AM structures, designers must consider not only how to optimize geometries but also how the geometrical designs influence the manufacturing qualities of parts.

It is generally recognized that with improved manufacturing flexibility, it becomes easier for the designs to focus on maximizing functionality by applying the design for functionality (DFF) rules. However, traditionally most research and development efforts in AM have been focused on new materials and process development, and much of the existing knowledge body is built upon empirical principles and experimental observations. As a consequence, many AM design guidelines are highly machine-specific or material-specific, and the guidelines often fail to provide comprehensive information set that is required to set up the manufacturing production correctly. For example, the optimum process parameters for an Inconel 718 turbine blade fabricated on an EOS M270 platform may not result in best fabrication qualities in a newer EOS M290 platform due to the improved inert gas flow control in the later system even though both are developed by the same manufacturer. Also, for many of these processes, there exist various input parameters that are not yet properly controlled or even specified. One such example is the powder spreadability of the powder bed fusion AM processes, which is closely related but different from the specifications of powder from traditional powder metallurgy. Although good flowability generally corresponds to good spreadability, the opposite conclusion does not always hold. The lack of generality with most current AM design guidelines makes it difficult for the adoption of new processes, new materials, or new product designs, which in turn has become one of the major barriers of the widespread application of AM in general industries.

Extensive studies have been carried out that use simulation based process modeling in the attempt to reveal comprehensive insights into various AM processes. However so far this approach has had limited success with metal AM processes, which is largely contributed by the inhibitive computational costs required for comprehensive part simulations. For metal powder bed fusion AM processes, there exist over 20 process input parameters that could potentially affect the quality of the fabricated parts, and possibly more that are not adequately identified. In addition, due to the rapidly evolving small-scale melting–solidification process, the resolution required to adequately represent the processes is often high. As a result, the simulation of the full-scale process often becomes exceedingly computational intensive [10].

For many traditional manufacturing such as machining, the mechanical and physical properties of the processed materials are normally standardized. Furthermore, during the manufacturing processes these properties also remain largely constant. On the other hand, for many AM processes and especially metal AM processes, the properties of the materials are often dependent on the geometrical designs of the structures. Therefore, geometrical design is an integral part of the design for additive manufacturing (DFAM) theory. In fact, the focus of development for DFAM has been gradually shifting from process focused guidelines to more integrated process-geometry design guidelines in recent years.

Unlike traditional design for manufacturing (DFM) theory, the DFAM is driven by design functionality, which in turn drives the material and process selection as well as geometrical designs. Currently, most existing theories with DFAM focus on either material/process optimization or structural optimization, and there are limited literatures that integrate both components. On the other hand, many qualitative guidelines can still provide valuable insights into the design of AM parts despite the lack of accuracy, which is especially valuable during the early design stages. In addition, such knowledge could also help companies to make decisions about the adoption of AM for specific purposes.

5.2 Material Considerations

Overall the material selection for AM is still relatively limited. Although this is partly caused by non-technical reasons such as market demand and costs, it often results in additional design difficulty as the available materials may not perfectly match the application requirements. On the other hand, the limited material/process selection does not simplify the decision making process since the material properties are process and geometry dependent. The material specifications provided by the system manufacturers only provide baseline references, and often requires additional scrutiny during applications.

Currently AM processible materials range across all basic categories of materials including metals, polymers, ceramics, and composites. Table 5.1 lists some of the commonly available metal materials for powder bed fusion AM and their specifications provided by the manufacturers. It can be seen that most of the listed metal materials are supported by most powder bed fusion AM systems, although there also exist materials that are only supported by certain systems, such as titanium aluminide (TiAl), which is currently only available via electron beam melting systems. In principle there should be very little difference of the material compatibility among different powder bed fusion AM systems. However, some of the systems appear to support more material selections likely due to business considerations.

The metal powder used by the powder bed fusion AM systems are similar to the powder used in some traditional manufacturing processes such as powder metallurgy and metal injection molding. However there also exist significant differences, especially when complete powder melting–solidification takes place during the process (e.g. laser melting and electron beam melting). Beside the traditional

Table 5.1 Typical metal materials for powder bed fusion AM

Metal material	Modulus (GPa)	Ultimate strength (MPa)	Elongation (%)	AM systems
Ti6Al4V [11, 12]	110–120	930–1020	10–14	Arcam, EOS, Renishaw, SLM Solution, 3D Systems, Concept Laser
CP-Ti [11]	–	570	21	Arcam, EOS, Concept Laser
316L Stainless steel [12, 13]	184–185	633–640	40	EOS, Renishaw, SLM Solutions, Concept Laser
Maraging steel [12, 14]	160	1110	10–11	EOS, Renishaw, 3D Systems
17-4PH Stainless steel [12, 14]	160–170	850–1300	10–25	EOS, SLM Solutions, 3D Systems, Concept Laser
Co-Cr-Mo [11, 12]	191–200	960–1100	20	Arcam, EOS, Renishaw, SLM Solution, 3D Systems, Concept Laser
IN625 [12, 13]	170–182	827–961	35	EOS, Renishaw SLM Solutions, Concept Laser
IN718 [12, 13]	166–170	994–1241	18	EOS, Renishaw SLM Solutions, Concept Laser
AlSi10Mg [12, 14, 15]	60–78	240–361	5–20	EOS, Renishaw, SLM Solutions, 3D Systems, Concept Laser

specifications such as powder particle size distribution, average particle size, and flowability, additional factors such as powder size distribution modes, powder spreadability and powder chemistry also need to be considered. Currently AM metal powder still consist only a small fraction of the total metal powder market, which affects both the cost and the development of new materials by powder manufacturers.

It should also be noted that the specified properties listed in Table 5.1 should only be used as a general guideline instead of design inputs, since the specifications were defined from standard testing samples fabricated via "standard" processing conditions that are specific to each system. As a rough rule of thumb, for most traditional designs the metal materials fabricated via AM should possess quasi-static mechanical properties comparable to or slightly better than the wrought parts, which is generally believed to be contributed by the refined microstructure. Even though the actual property values can be difficult to predict, this rule can be helpful in providing baseline expectations for process development. On the other hand, some mechanical properties for AM metal materials such as fatigue strength and creep strength that are critical to many applications are much less well understood, and the lack of baseline references with these properties often hinders efficient process development and design selection.

Some AM processes such as directed energy deposition and binder jetting support much wider range of metal materials. For example, various grades of titanium alloys have been used successfully in the laser based directed energy deposition processes [16], and exotic materials such as tungsten, niobium, and molybdenum are offered for the wire-fed electron beam directed energy deposition systems [17]. In principle binder jetting AM is capable of handling almost all the powder materials as long as there exist compatible binder systems, therefore appears to provide a desirable solution for material selection issue often encountered with other AM processes. However, due to the need for secondary processes in order to achieve satisfactory mechanical properties, binder jetting AM is subject to additional design constraints compared to many other AM processes. In addition, despite various demonstrations of the material capabilities, relatively few metal materials have been adequately developed on binder jetting platforms for application purposes.

Various polymer materials have been used by AM processes such as vat photopolymerization, material jetting, material extrusion, and selective laser sintering. As one of the most versatile types of materials, polymers can be tailored to have desired combination of properties, therefore possess great potentials in the design of advanced materials/structures with AM. Compared to traditional polymer manufacturing processes such as injection molding and thermoforming, AM polymer processes generally have significantly lower production rate (material mass per unit time). Additionally, the relatively limited selection of commercially available AM polymer materials is a limiting factor for performance-driven designs. As a result, the applications of polymer AM technologies has been comparatively limited. Many of the current applications with AM polymers have taken advantage of various physical properties of polymer materials such as low density, chemical resistance, optical transparency and electrical insulation. However, more polymer materials and polymer matrix composite materials are being rapidly developed that could potentially be used for high performance end products, with the carbon fiber reinforced polymer matrix materials [18], high strength PEEK and PEKK [19] being some of the examples. Table 5.2 shows some of the most commonly available AM polymer materials and their specifications provided by the material manufacturer. It should be noted again that most of these property values should only be used as references rather than design inputs, since additional design and manufacturing factors will also affect the properties of the fabricated structures. For example, even though ABS exhibits properties that might be adequate for certain applications, during the material extrusion processes, due to the highly directional material deposition patterns, the final structures will exhibit significant anisotropic properties and lower performance along the layer stacking direction (Z-direction). In addition, due to the typical variability of property for most polymer materials, it is also more difficult to apply AM polymer to structural designs with stringent performance requirements.

Compared to the other two types of materials, ceramic materials are much less used in AM. Only a few types of ceramic materials are currently used on the commercial AM systems for specialized purposes. Table 5.3 lists some of these ceramic materials and their common applications. Similar to the issues encountered with traditional processes, the manufacturability of ceramics makes them challenging for most AM processes. On one hand, ceramic materials often possess the combination of high melting temperature, high coefficient of thermal expansion and low fracture toughness, which makes it extremely challenging for the melting-solidification process approach such as the one taken by powder bed melting processes. On the other hand, when a more traditional green part-densification approach is taken, the design will be subject to manufacturability limitation similar to the traditional ceramic processes such as thermal inhomogeneity, debinding, sintering distortion, and porosity. Lack of recognition of suitable applications is yet another limiting factor for broader adoption of AM ceramics. Currently the commercial ceramic AM systems, utilizing the binder jetting, direct write and photopolymerization processes,

Table 5.2 Typical polymer materials for AM

Material	Glass transition Temp. (°C)	IZOD notched impact strength (J/m)	Modulus (MPa)	Tensile strength (MPa)	AM platform
Nylon 12 [20, 21]	176	32–42	1700	50	Laser sintering
PEEK [21]	–	–	4250	90	Laser sintering
ABS [19]	108	96.4–128	1920–2200	28–37	Material extrusion
PLA [22]	56–57	16–47	2020–3544	49–56	Material extrusion
Ultem 9085 [19]	186	48–120	2150–2270	42–69	Material extrusion
Epoxy and Acrylate based photopolymer [20]	56–58	12–25	2690–3380	58–68	Photopolymerization, material jetting

Table 5.3 Typical ceramic materials for AM

Material	Melting point (°C)	AM systems	Applications
Silica sand [23]	1290	Binder jetting—ExOne, Voxeljet	Sand casting
Calcium sulfate hemihydrate (plaster) [24]	1450	Binder jetting—3D Systems ProJet	Design prototype
Alumina [25]	2070	Direct write/Photopolymerization	Prototype, labware, R&D
Zirconia [25]	2715	Direct write/Photopolymerization	Prototype, labware, R&D
Tri-calcium phosphate [25]	1391	Direct write/Photopolymerization	Prototype, tissue scaffolds, R&D
Hydroxylapatite [26, 27]	1670	Photopolymerization	Tissue scaffolds, R&D

all take the second fabrication approach and focus on the realization of geometries. Depending on the applications, secondary densification processes such as sintering or infiltration are often needed in order to achieve the final strength of the structures. However, for the ceramic structures with complex geometries shown in Fig. 5.3 that is unique to AM technologies, additional challenges for the sintering densification processes remain little understood, which represent only one of the various gaps between the technological capabilities and the end-use requirements. Most of the current AM ceramic technologies have demonstrated the final part densities of about 95% after densification, which still poses significant challenges for their potential use in structural applications due to the high sensitivity of defects with ceramic materials. There also exist applications where green parts can be directly used. For example, sand molds and cores fabricated via binder jetting systems (e.g. ExOne, Voxeljet) are used directly for the sand casting production. With binder jetting processes, standard foundry sands can be used directly as long as the spreadability of the powder meets the requirement of the particular system. By using various foundry binders such as furan binders, phenolic binders and silicate binders, the quality of the printed molds and cores can be comparable to the traditional ones, although additional design consideration must be taken in issues such as mold permeability and mold accuracy.

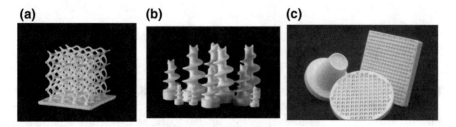

Fig. 5.3 Ceramic parts. **a** Lattice structure [28]. **b** Custom screws [28]. **c** Filtration [29]

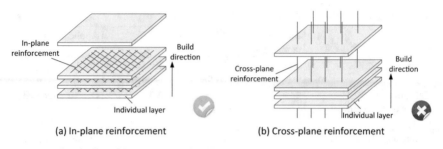

(a) In-plane reinforcement   (b) Cross-plane reinforcement

Fig. 5.4 Continuous reinforcement with AM technology

Composite materials are in general more challenging for the current AM technologies. As shown in Fig. 5.4, due to the layer-wise process of AM, the discontinuity in the build direction (Z-direction) does not allow for continuous reinforcement in that direction. Instead, continuous reinforcement can only be realized within the build planes, which is sometime undesired as this would further increase the anisotropy of the structures. Currently, AM composites exist mostly in the form of particle reinforced composite structures through addition of particulate reinforcement phase in the primary materials. One such example is the glass-filled nylon powder for powder bed sintering processes for improved impact strength and rigidity. There also exist AM systems that fabricate continuous fiber reinforced composite structures, which is suggested to have in-plane properties superior to that of the regular aluminum alloys [18].

It can be seen that it is logical for many designers to take design approaches similar to the traditional material-first process design approach, since there only exist limit selections for each type of materials. However, it should be emphasized again that the properties of the AM materials are dependent on various factors including process parameters, build orientations, geometrical designs in addition to the material feedstock related factors. Also, from functionality-driven design perspective, the properties of the material should be simultaneously designed as the geometries are designed instead of being separately defined. With the current DFAM theory, it is still difficult to realize such design approaches, however, with rapid development in this area, combined with the ever-growing industrial adoption of AM technologies, it can be expected that this knowledge gap will be quickly filled up.

5.3   General Design Consideration for AM

Although each AM processes exhibit unique characteristics, there also exist issues that are common to most AM systems. Due to the layer-wise process, the geometrical qualities (e.g., feature resolution, surface finish, and geometrical accuracy) of AM systems along the build direction (Z-direction) are usually affected by the

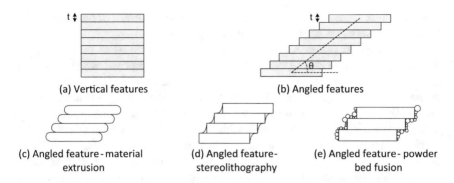

(a) Vertical features   (b) Angled features

(c) Angled feature - material   (d) Angled feature-   (e) Angled feature - powder
        extrusion   stereolithography   bed fusion

Fig. 5.5 Staircase effect of AM parts

layer thickness. While some metal powder bed fusion AM systems offer layer thickness of as small as 20 μm, other system such as stereolithography and material jetting can readily achieve finer thickness of at least 16 μm [30, 31]. Although small layer thickness generally indicates higher geometrical accuracy and feature resolution, there exist various other factors that play important roles in determining the geometrical qualities of a part. For vertical features (Fig. 5.5a), the staircase effect caused by the layer-wise process is minimum, and the geometrical accuracy in Z-direction is determined by both the layer thickness and the shaping characteristics (e.g., deposition profile, shrinkage, etc.) of the materials during the process. On the other hand, for angled features (Fig. 5.5b), the staircase effect is determined by both layer thickness and the feature angles. Smaller feature angle results in not only more significant staircase effect but also less effective bonding length between layers, which could in turn affect the mechanical properties of the structures. From Fig. 5.5b, the roughness of the side surfaces can be roughly demonstrated by applying geometrical analysis as:

$$R_a = \frac{t \cos \theta}{4} \tag{5.1}$$

where t and θ are the layer thickness and feature angle respectively as shown in Fig. 5.5b. It should be noted that Eq. (5.1) only serves to demonstrate the staircase effect, while in reality more complex analysis is needed in order to accurately estimate the surface quality. For different processes, the staircase effect exhibit different characteristics such as shown in Fig. 5.5b–e. For material extrusion processes, due to the shaping effect of the extrusion nozzle and the gravity-induced distortion, the deposited tracks exhibit round ends and flattened cross section profiles (Fig. 5.5c); For stereolithography processes, due to the surface wetting between the cured part and the uncured resin, additional resins tend to attach to the stairs and consequently cured during the post-processes, which helps to smoothen the side surface (Fig. 5.5d) [32]; For powder bed fusion processes, the surface sintering effect causes additional powder to attach to the surface of the parts, which increases the variation of surface qualities (Fig. 5.5e).

For powder bed fusion AM, another factor that contributes to the geometrical error is the shrinkage during the melting–solidification process, which is in turn largely determined by the powder bed density. The achievable packing density for the powder bed is largely determined by the morphology of the powder including shape, size distribution, as well as surface characteristics of the powder. As a general rule of thumb, powder particle with more regular shapes can achieve higher packing densities. While very irregular powder can only achieve relative density of as low as 40%, the highest theoretical relative density for mono-sized spherical powder bed is about 65% (Fig. 5.6) [33]. Although in theory higher relative density of around 82% can be achieved via mixing powder with different particle size distributions, the addition of smaller particles tend to increase the interparticle friction and therefore reduces the flowability of the powder [34]. Therefore, for practical purposes, the relative density limit of the powder bed is 68–70%. The porosities included in the powder bed will result in volumetric shrinkage during melting/solidification, which reduce the geometrical accuracy in all directions. For isotropic shrinkage, such volumetric shrinkage would corresponds to ~10% of linear shrinkage which is considerable for the design purposes. However, for most AM processes, the solidification shrinkage occurs anisotropically, with the shrinkage in the build plane directions restricted by the interaction between the current layer and the previous layer/substrate. For powder bed fusion processes, the error in Z-direction does not have accumulative effects since for each additional layer the newly added layer thickness will always be filled with powder and therefore compensate for the shrinkage. However, the same conclusion could not be made for the other processes. For example, for material extrusion processes, the feature resolution and accuracy in Z-direction the once the printing parameters (e.g. extrusion speed, printing speed, temperature) are fixed, the amount of material deposited on each location is also fixed. Therefore, geometrical errors could

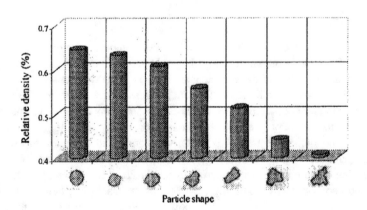

Fig. 5.6 Achievable relative density with different particle shape [33]

accumulate over multiple layers, which could potentially result in significant loss of accuracy for the parts if not accounted for carefully.

For many AM processes, the feature resolution of a part is closely associated with its geometries. In another word, for these processes, there does not exist a standard process resolution that can be applied to all designs. This can be quite frustrating during the design processes, especially when the detailed geometries are still subject to changes. Various works have been attempted to establish standard benchmark parts with sets of predesigned features in order to provide estimations of the geometrical qualities of specific AM systems, but in general the success has been limited. So far most benchmark parts can be categorized into three types [35]:

a. Geometric benchmark: used to evaluate the geometrical quality of the features generated by a certain machine.
b. Mechanical benchmark: used to compare the mechanical properties of features or geometries generated by a certain machine.
c. Process benchmark: used to develop the optimum process parameters for features and geometries generated by certain process systems or individual machines.

The three types of benchmark parts serve different design evaluation purposes. The geometric benchmark evaluates geometrical characteristics of a certain material/process parameter combination including accuracy, precision, surface finish, and repeatability, which can be either qualitative or quantitative. Geometric benchmark parts usually incorporate multiple geometrical features that are representative to basic types of traditional feature-based geometries such as blocks, cylinders, domes and prisms, and geometrical characteristics and tolerances (GD&T) will be measured on these features to provide evaluations. So far most of the proposed AM benchmark parts fall into this category. Figure 5.7 shows some of the proposed designs for geometric benchmark. This type of benchmark part can be effective in optimizing process selection and process parameters for a fixed design, since the geometries used for evaluation are the final geometries. On the other hand, GD&T information generated by these geometric benchmark parts could not be easily generalized, therefore lacks the insight into the intrinsic process characteristics.

Mechanical benchmark parts such as mechanical testing coupons focus on evaluating the quantitative mechanical properties of a part under certain material/process combinations. Mechanical benchmark studies generally face similar challenges that the generated knowledge could not be easily generalized. Without systematic design guidelines, currently the mechanical properties of the AM parts are mostly evaluated via standard testing coupons following existing material characterization standards (e.g. ASTM E8, ASTM B769), which has been arguably one of the most misleading approaches in the understanding of AM characteristics. For most AM processes, there exist a correlation between the properties and the characteristic dimensions of a part. For example, for many material extrusion processes, there exist separate deposition patterns for the boundaries and interiors of the 2D areas as demonstrated in Fig. 5.8a. Since the boundaries and interiors are treated with different deposition patterns, which brings

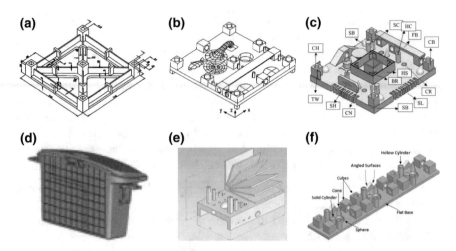

Fig. 5.7 Geometric benchmark design proposals. **a** 3D Systems [36]. **b** Childs and Juster [37]. **c** Mahesh et al. [38]. **d** Kim and Oh [39]. **e** Castillo [40]. **f** Fahad and Hopkinson [41]

about different properties to different regions of the structures, this treatment essentially creates shell-core composite structures. Therefore, when the characteristic dimensions of the part changes, the percentage of boundary versus interior regions could potentially change, which results in the change of mechanical properties. Some material extrusion systems provide software that allows for complete control of the deposition strategies by the users, while some systems focuses on providing a "turn-key" solution and therefore do not open such adjustability to users. Another example comes from the powder bed sintering processes. Currently in the laser sintering systems the scanning deflection of the laser is realized via the control of a scanning mirror. As the scanning operation requires frequent acceleration and deceleration of the motion of the scanning mirror, at the beginning and end of a scanning vector the laser often experiences a kinematic dwelling effect called the end-of-vector effect, which could potentially overheats the scanned areas. Similar issues could also be observed for smaller areas that have smaller scanning vectors and therefore requires more frequent scanning "turning" controls. As a result, even under identical process conditions, the variation of dimensions of a part could cause regions with shorter scanning vector length to exhibit defects as shown in Fig. 5.8b.

For powder bed melting, the size dependency issue is even more significant. This is due to the fact that the solidification and microstructure of the metal materials are very sensitive to the thermal conditions and heat transfer processes. In metal powder bed melting process, the energy source moves along specific patterns at high speed, which creates rapidly moving thermal gradients. It is obvious that the choice of scanning strategies affect the temperature field characteristics.

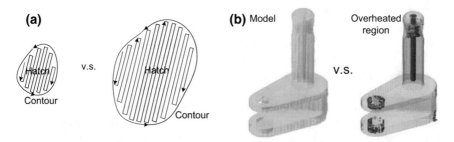

Fig. 5.8 Size-dependent characteristics of AM parts. **a** Material extrusion deposition pattern. **b** Laser sintering scanning vector length effect [42]

For example, with simple zig-zag scanning, the temperature gradient in the part tends to increase with increased scanning vector length. This is due to the fact that longer scanning vector lengths results in longer interval between two consecutive heating of the same location. Increased temperature gradient could potentially result in faster cooling and supercooling with the melting pool and therefore different microstructure. In addition, since the thermal conductivity of powder bed is significantly lower compared to solid material, during the powder bed melting processes the primary heat conduction path goes downwards through the previously fabricated structures, which also means that the previously fabricated structures could also be subject to additional heating–cooling cycles, which potentially act as heat treatment that alters the microstructure and properties of the structures.

It is worth noting that the thermal gradient distribution issue also affects the geometrical accuracy of the metal powder melting AM parts. Larger thermal gradient often results in larger thermal residual stress and therefore distortions if not countered adequately, which affects the geometrical accuracy of the parts. Although most metal powder bed AM systems employ sophisticated scanning strategies in order to reduce thermal residual stress, for parts with varying feature sizes, it is still difficult to accurately predict the thermal residual stress and design for optimized processes accordingly. For powder bed fusion AM processes, sometimes it is fine to have positive dimensional errors with the fabricated parts, since the extra materials can be removed by the secondary surface treatments that is needed to achieve improved surface finish.

Mechanical benchmark parts and geometric benchmark parts often serve as process benchmark parts at the same time, since many process development aims to optimize either geometrical or mechanical properties of the parts. Therefore it can be easily seen that process benchmark parts faces similar challenge as the other two types of benchmark parts such as the property dependence on the actual geometry of interest. In the cases where the input parameters have compound effects on the material properties and geometrical characteristics, even qualitative evaluation can become difficult. A compromised approach that could be useful is to categorize the features of interest into representative groups, and consequently perform design of experiment with each group, which is more likely to exhibit regular trends and patterns in response to the change of input parameters. It is sometimes equally difficult to effectively define the categories, but there also exist some commonly recognized design input variables that likely need to be investigated by experimental designs. Table 5.4 lists these parameters and their potential effect on geometrical and mechanical characteristics of parts. By taking a subset of these parameters and focus on the experimental investigations, it is possible to obtain knowledge efficiently for a certain generic type of geometries (e.g. blocks, cylinders, thin-wall features) and optimize processes.

National Institute of Standards and Technology (NIST) recently proposed a standard benchmark part that aims to provide both geometric and process benchmark [44]. As shown in Fig. 5.9, the part consists of multiple features that a represent relatively comprehensive set of GD&T characteristics, and some of the feature designs also considered unique issues with AM such as overhanging features, minimum feature sizes and extrusion/recession features. On the other hand, it does not include features for angular and size dependency evaluations. This benchmark part was primarily designed for powder bed fusion AM processes, while other designs might be potentially better suited for other AM processes.

Another design issue common to most AM is the design for quality issue. The quality control in AM is currently realized via two ways: in-process process certification/measurement and post-process part qualification. For traditional manufacturing processes, both in-process quality control and post-process part qualification have been extensively employed, and the parts can be designed to maximize the quality stability of the processes accordingly. However, for AM processes, feedback closed-loop process control is generally unavailable except for a few commercialized systems such as the DM3D directed energy deposition system [45]. For some AM systems such as powder bed fusion systems, this is not only caused by the difficulty in setting up effective in-process sensing system but also the lack of understanding of the relationships between the input control variables and the measurable output variables such as temperatures and thermal distortions. Currently, most AM manufacturing systems are operating without closed-loop control. The quality control for an open-loop AM manufacturing system for production often requires complete fixation of all input variables including material feedstock supplier and specification, process system, process parameters, part setup, post-process and maintenance. Even with these measurements, due to the random quality variations of the material feedstock and the potential variations of unspecified process control parameters (e.g. powder bed surface lateral air flow rate), the quality of the final parts could still exhibit considerable variations. In some literatures, real-time information such as temperature history acquired through in-process sensing are correlated to the post-process inspection results in order to establish empirical process–quality relationships, which shows initial promises but still requires extensive further development [46–49]. On the other hand, the post-process measurement of AM parts can become difficult especially for freeform geometries and parts with support structures attached. For the parts shown in

Table 5.4 Common AM process development input variables

Input parameters	Geometrical effects	Mechanical effects
Feature angle	Larger angle in relation to the Z-direction corresponds to lower accuracy and higher roughness on the sides	Most process exhibit anisotropic properties. In addition, mechanical property-angle relationships are not all similar. For example, for powder bed melting AM, fatigue strength and tensile strength may have opposite angular trends [43]
Extrusion/recession features	Recessed features can generally achieve smaller minimum feature sizes. Recessed features may be clogged. Recession and extrusion features are subject to completely different heat transfer conditions during the processes	Recession and extrusion features are subject to completely different heat transfer conditions during the processes
Overhang	Larger overhanging angle or larger overhanging area generally corresponds to more significant loss of accuracy. Down-side facing surfaces of the overhanging features usually have more rough finish	Overhanging features are subject to different heat transfer conditions as the downward heat transfer is restricted
Cross-sectional dimensions	Larger cross section may indicate larger heat accumulation during process and larger heat sink afterwards, which affects thermal distortion	Larger cross section may result in more significant in-situ heat treatment effects due to the heat accumulation and heat sink effect. For processes that adopt contour-hatch fabrication strategies, different dimensions may result in different combination of deposited properties
Change of cross-sectional area dimension	Similar to cross-sectional effects. Change of dimensions may result in additional stress concentration due to the change of heat transfer conditions, which may cause additional thermal distortions	The change of dimensions may result in change of microstructure and properties
Input energy density	Input energy may affect how the materials are fabricated. For example, insufficient energy input may cause melting pool instability in powder bed melting AM processes and therefore affect the part accuracy and surface finish	There often exist a window of input energy density that results in relatively optimal mechanical properties

(continued)

Table 5.4 (continued)

Input parameters	Geometrical effects	Mechanical effects
Layer thickness	Smaller layer thickness usually corresponds to higher part accuracy at the cost of manufacturing time	For some AM processes, layer thickness indicates the thickness of a "laminated composite" structure, therefore is closely associated with the properties. In some processes individual layers serve as boundaries of microstructural grains. Smaller layer thickness may result in better side surface finish which is beneficial to fatigue performance
Material feedstock density	Lower material feedstock could result in more shrinkage during the process. For some processes, lower feedstock density also indicates higher process instability, which reduces geometrical accuracy	Higher material feedstock density often corresponds to higher overall mechanical performance due to the reduced amount of internal defects
Preheat	Preheat may cause additional material distortion or loose powder attachment, which reduces accuracy. On the other hand, preheat reduces thermal distortion and therefore could be beneficial to the overall part accuracy	Preheat affects thermal gradients in the processes, which in turn affect the microstructure. Preheat could sometimes serve as in-situ heat treatment and alter the microstructure

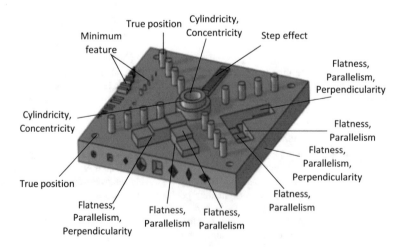

Fig. 5.9 NIST proposed standard benchmark part design

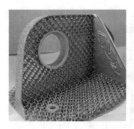

Fig. 5.10 Some freeform AM parts that are difficulty to inspect

Fig. 5.10, without effective methods to perform inspection to individual features, even the establishment of quality evaluation protocol can be nearly impossible.

5.4   Support Structure Designs

Support structure is an integral part for many AM processes, and the design of support structure is no trivial task. On one hand, support structure is necessary for many part designs, and on the other hand, support structure is generally undesired. The fabrication of support structure usually cause additional time and materials, and the removal of support structures not only requires additional processes, but also poses additional design constraints to the parts.

Support structures are needed for various reasons. For vat photopolymerization, due to the existence of the buoyancy force and the shrinkage induced distortion effect during the photopolymerization processes, the support structures are needed to ensure the quality and accuracy of the parts. For material extrusion processes, both thermal residual stresses and gravity necessitate the need for support structures. For powder bed fusion processes, the main purpose of the support structure is to counter the thermal residual stresses generated during the melting–solidification processes. For metal powder bed fusion processes and especially the laser based systems, the large temperature gradients that are generated during the fabrication process often require extensive support structures that are carefully designed.

Various AM preprocessing software support automatic support structure generation, which is largely based on the geometrical characteristics of the parts. Figure 5.11 shows several situations where support structures are usually generated for. For many AM processes, a general guideline is that features with overhanging angle of over 45° require support structures (Surfaces C and D in Fig. 5.11a), although this condition can be relaxed under certain circumstances. For example, when it is acceptable to have rougher surface finish for the down-facing surfaces, overhanging angle of up to 70° is still realizable if the process parameters are set up properly [50]. In addition, features that are unsupported (cup handle Fig. 5.11b) and overhanging features with very large projected areas (Fig. 5.11c) generally require support structures. The algorithms for the support structure generation calculate the unsupported projected areas of each layer, and determine the need for support structures based on various conditions such as the total areas, area change over the layers and interference with the parts. For some special geometries, such as the concave and convex filet features shown in Fig. 5.12, even though the features possess large overhanging angle and projected overhanging areas, due to the gradual transition of features, the overhanging can be adequately self-supported with proper process designs [51]. Various types of support geometries are available from multiple support generation software, such as the point support, web support, line support, block support and contour support shown in Fig. 5.13 [52]. In addition, the contact between the support structures and the parts can be either points, lines or areas depending on the required support strength. Point contact provides the weakest attachment strength, but is also the easiest to remove in the post-process. On the other hand, area contact generally provides the strongest attachment strength, but could pose significant challenges for the support removal process. Support removal can be facilitated by applying different materials for support structures that are easy to be removed through mechanical, physical or chemical means such as heating or dissolution. For processes that fabricate parts and support structures using the same materials, which include all the current metal powder bed fusion processes, such scenario would not be applicable.

For the support geometries shown in Fig. 5.13, there exist some general guidance. For example, block support is usually used for bulk geometries, while point and line supports are used for small features. Contour support can be considered when the contours of the parts need to be better sustained. However, little decision

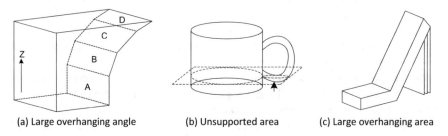

(a) Large overhanging angle   (b) Unsupported area   (c) Large overhanging area

Fig. 5.11 Typical geometries that require support structures

Fig. 5.12 Overhanging features with transition that may be self-supportive

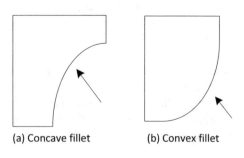

(a) Concave fillet   (b) Convex fillet

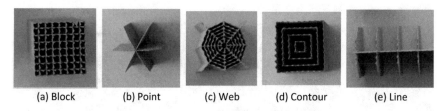

(a) Block   (b) Point   (c) Web   (d) Contour   (e) Line

Fig. 5.13 Typical support geometries for metal powder bed fusion AM [27]

support is available for the detailed design of these supports such as the support spacing and the contact area, and the design processes are still largely based on the designers' personal experiences. In many cases, in order to ensure the success of fabrication, the support structures are often over-designed, which often imposes significant challenges for the support removal and turns this process into a hand-crafting type of work. For example, the support structures shown in Fig. 5.14 may require a series of manual operations such as cutting, machining, grinding and polishing in order to meet the end use specification, and the complex geometries of these support structures as well as their interfaces with the parts poses various design for manufacturing issues to these processes.

Based on the geometries of the fabricated structures, it is often possible to reduce the amount of support structures required via the adjustment of orientation. However, the orientations that result in minimum amount of support also often result in very long fabrication times, since under these cases the largest dimensions of the parts are more or less aligned to the build direction, such as the example shown in Fig. 5.15. Another factor to consider during the determination of part orientation is the ease of support removal. Since the removal of support often damages the surface, critical surfaces with requirement of integrity and surface finish should avoid support structures. Also, support should be avoided for features such as surfaces with small clearance, internal channels and visually obstructed

(a) A biomedical implant   (b) A metal joint part

Fig. 5.14 Complex support structures generated for laser powder bed fusion parts

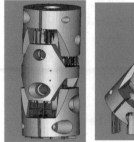  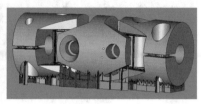

(a) Fewest support   (b) Easy-to-remove support   (c) Shortest fabrication time

Fig. 5.15 Support generation by Materialise Magics for a universal joint

surfaces. As a result, for complex geometries as illustrated in Fig. 5.15, the optimal part orientations for support removal are often neither horizontal nor vertical but instead intermediate angles in between.

In the detailed design of support structures, several variables are generally considered to be of importance. As shown in Fig. 5.16, for block support, the hatch spacing, or the space between neighboring support lines, determines the density of the support, and generally smaller hatch spacing corresponds to stronger support effect and smaller thermal residual distortion. For individual support bars, teeth structures are usually designed to reduce the total contact between the support and the part and to facilitate the support removal. Various design variables such as tooth height, tooth spacing, offset, base interval, base length and top length as shown in Fig. 5.16b can be considered. Top length was shown to be a significant variable, and larger top length generally corresponds to stronger support effects [53]. Larger offset could also help improve the strength of the support attachment, although this effect diminishes with larger top length. On the other hand, since support structures and parts are usually fabricated with different process parameters, the overlapped regions resulted from the offset setting are more prone to microstructural defects. Also, smaller tooth spacing tend to strengthen the support effect since it corresponds to increased number of teeth.

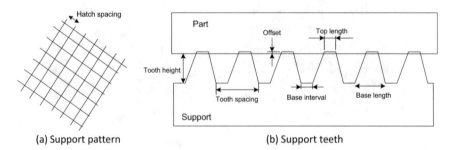

(a) Support pattern   (b) Support teeth

Fig. 5.16 Typical design variables for block support

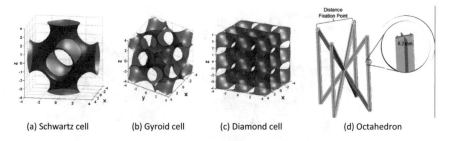

(a) Schwartz cell   (b) Gyroid cell   (c) Diamond cell   (d) Octahedron

Fig. 5.17 Several unit cell geometries for support structures [50, 54]

In the attempt to further improve the material efficiency and facilitate the ease of removal of the support structures, an emerging trend is to apply cellular structures to the support designs. Compared to the traditional thin-wall support structures, cellular support structures could potentially achieve better material efficiency due to the 3-dimensional porosities with their geometries. In addition, the more porous cellular support structures are also expected to facilitate the support removal processes. As shown in Fig. 5.17, different cellular geometries can be adopted for support design, although currently no theory is available to guide the selection of these geometries. Based on the traditional design theory of cellular structures, the mechanical properties of the cellular supports can be controlled via the design of relative densities, which in turn can be realized rather straightforwardly via the change of cellular beam/wall thickness or cellular unit cell sizes. Experimental-based design evaluation and optimization can be carried out to determine cellular designs with adequate support strength, since numerical methods could potentially suffer from the inhibitive computational loads. Similar to the traditional support structures, the support effect of the cellular supports are also significantly influenced by the number of contact points with the parts. In comparison to the traditional support designs, the cellular support design may encounter more significant "sagging" issues at the support-part interfaces due to the limited number of contact points in given areas. One proposed solution is to create sacrificial transition solids between the cellular supports and the actual parts as shown in Fig. 5.18 [50]. The transition layers can be fabricated with process parameters that are optimized for minimum thermal residual stress and best surface finish, so they serve as satisfactory substrate for the actual parts. Other factors that need to be determined for economical purposes are the fabrication time and weight efficiency for each type of cellular geometries. Due to the fragmented cross sections in each layer, cellular structures often require significantly extended time to scan for most laser based powder bed fusion AM systems. On the other hand, for electron beam based systems, due to the capability of electron beam control system to realize inertia-free high speed scanning deflection, the fabrication time is not as significantly affected by the geometry fragmentation.

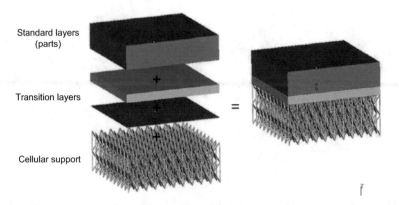

Standard layers
(parts)

Transition layers

Cellular support

Fig. 5.18 Transition layer design for cellular supports [50]

Various commercial AM software including Materialise Magics and Netfabb readily offer the capability to design and visualize these cellular geometries from unit cell libraries, therefore makes it possible for the broader utilization of this approach.

Since for powder bed fusion AM the primary function of the support structures is to counter thermal residual stresses generated during the fabrication processes, the correct evaluation of thermal residual stress is of critical importance to the successful design of these supports. From this perspective, it is also obvious that the geometry-based support generation algorithms could not satisfy the design requirements. Due to the computational difficulty of full-scale simulation, few results are currently available in the guidance of thermo-mechanical characteristic based support design. In one such works, a fractal support generation approach was proposed that aims to realize localized design support optimization based on thermal residual stress calculations. As shown in Fig. 5.19, the support areas can be divided into higher order subdivisions as needed, and during the iterative simulation based designs, if the thermal residual stresses of a certain unit area exceeds the designated threshold, then a higher order division of that unit area can be performed, which results in strengthened support effects in that unit area. Recently, 3DSIM LLC announced the release of a support generation optimization software that is expected to be integrated to the full-scale process simulation software the company is currently developing, which may have the potential to provide users with such design capabilities [55].

Another interesting effort from the commercial side was reported by Autodesk, who is experimenting with a newly developed organic branch-like support generation algorithm. As shown in Fig. 5.20, using a software named Meshmixer, branch-like support structures can be created with the objective of minimizing material use and maximizes support removability. Although this algorithm does not take thermal residual stress into consideration and therefore still lacks applicability to metal powder bed fusion processes, the concept of this approach may be further developed in more sophisticated support generation algorithms.

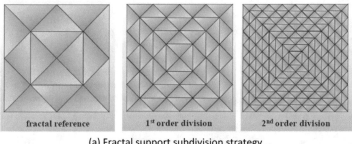

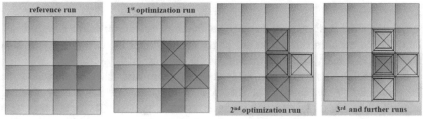

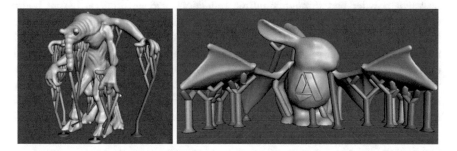

Fig. 5.19 Local support structure optimization based on thermo-mechanical simulation [52]

Fig. 5.20 Organic branch-like support generated by Autodesk Meshmixer [56]

5.5   Design Consideration for Powder Bed Fusion Metal AM

The resolution of the powder bed fusion AM processes are limited by various factors with energy beam profile and powder bed characteristics being the two most influential ones. The interaction between the energy beam and the powder bed creates localized melting pool, which experiences various phenomenon such as Marangoni convection and evaporation. In addition, the capillary forces at the boundaries between liquid phase and surrounding powder also acts as a potential for liquid phase permeation into the powder bed and the migration of powder

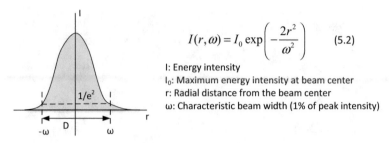

$$I(r, \omega) = I_0 \exp\left(-\frac{2r^2}{\omega^2}\right) \qquad (5.2)$$

I: Energy intensity
I_0: Maximum energy intensity at beam center
r: Radial distance from the beam center
ω: Characteristic beam width (1% of peak intensity)

Fig. 5.21   2D representation of Gaussian beam energy intensity distribution

particles. As the melting pool advances with the advancement of the energy beam, the dynamic evolution of the liquid phase could further increase the liquid–solid interaction instability and therefore affect the shaping quality.

It is generally assumed that both electron beam and laser beam follow the Gaussian energy distribution as shown in Fig. 5.21 with the energy intensity determined by Eq. (5.2), although it has also been suggested that for electron beam the radius distribution might deviate slightly from the ideal Gaussian shape [57]. When the energy beam is applied to the powder bed surface, part of the incipient energy is absorbed by the powder particle, while the remaining energy is carried away with the reflected energy beam. As shown in Fig. 5.22, the energy beam could travel in powder bed through multiple reflections and absorptions before being completely absorbed or escaping from powder bed. Due to this phenomenon, the total energy absorption rate (energy incoupling) for the powder bed is generally much than that of the solid material [58]. In addition, this also implies that the energy absorption rate for the powder bed fusion AM is powder batch specific, which sometimes necessitates experimental characterization when accurate energy absorption information is needed. Figure 5.23 shows some of the simulated results for the relationship between energy incoupling and solid material absorption for

Fig. 5.22   Energy absorption in powder bed

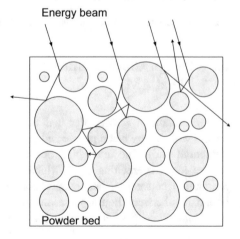

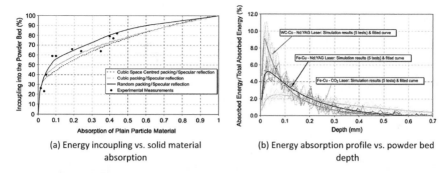

(a) Energy incoupling vs. solid material
absorption

(b) Energy absorption profile vs. powder bed
depth

Fig. 5.23 Simulated energy absorption characteristics for laser melting processes [58]

laser melting process, as well as the energy absorption profile in the depth direction of the powder bed [58]. It is obvious that the highest energy absorption occurs in the regions underneath the surface of the powder bed, which is in turn dependent on the powder bed characteristics such as powder morphology and particle sizes.

It has been mentioned previously that the powder morphology and size distribution could significantly affect the density of the powder bed, which in turn affects the energy beam–powder bed interaction. Although higher powder bed porosity might promote the overall energy incoupling, it also results in more dispersed distribution of energy, and therefore compromise the fabrication quality by reducing geometrical accuracy and introducing internal defects. It should also be noted that the selection of powder size distributions is also closely related to the powder spreading mechanisms in each individual systems. Multiple powder spreading mechanisms have been employed in various powder bed fusion AM systems, including roller, comb blade, sweep blade, and moving hopper. It is commonly perceived that the roller mechanism could potentially result in higher powder bed packing density that is beneficial for quality control of the processes. On the other hand, the additional shear and compressive stress exerted to the powder bed via rolling could potentially lead to loss of accuracies of the parts and even damage to the structures. This is especially significant when the surface is not perfectly flat due to either local defects or part warping, since the roller would interfere with the surfaces. In addition, for powder with relatively low flowability, roller mechanism also tends to be less effective in rendering uniform and complete powder spreading. On the other hand, mechanisms such as comb blade and moving hopper tend to exert less compacting effects on the powder bed, however are also more forgiving to surface imperfections.

In general, the powder spreading mechanisms move along a plane that is slightly above the workpiece surface, and the clearance is closely related to both the powder size distributions and the layer thickness setup. While the theoretical minimum layer thickness is limited by the mean particle size of the powder, which determines the smallest thickness dimensions possible that ensures complete filling of a new

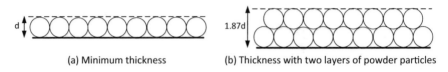

<table>
<tr><td>(a) Minimum thickness</td><td>(b) Thickness with two layers of powder particles</td></tr>
</table>

Fig. 5.24 Layer thickness versus powder particle sizes

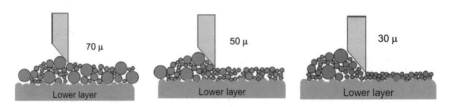

Fig. 5.25 Powder spreading with different layer thickness [59]

layer as shown in Fig. 5.24, it must be realized that for most commercial powder the particle sizes spread over a range, and consequently the powder spreading mechanisms might also serve as particle size filtering mechanisms that preferably spread smaller particles on the powder bed. As shown in Fig. 5.25, when larger particles are included in the spread layer, they tend to introduce larger voids, which in turn causes local shrinkage variation and consequent defects. On the other hand, if the layer thickness is small, then the spreading mechanisms could potentially move the larger particles out from the powder bed, which results in powder bed with smaller average particle sizes and better packing densities.

Although currently little systematic knowledge is available for the selection of optimal powder for specific part and process designs, various studies have shown that in general a bimodal distribution result in higher powder bed density and consequently better surface quality of the fabricated parts [34, 60–62]. A particle size ratio of 1:10 between the two mean particle sizes along with a percentage ratio of 10–15% for the smaller particles were suggested to yield optimal powder bed density and surface finish [61, 62]. It was suggested that powder with smaller particle sizes is generally more favorable for fabrication as long as it could be successfully spread, since it facilitates more efficient melting that is beneficial for achieving higher part density and surface quality [63]. However, the larger particles in the powder bed also appear to bring about some benefits such as higher elongation [60].

When choosing powder with single size distributions, the observations from different studies are less consistent. Narrower particle size distribution is generally considered to facilitate the powder flowability and spreadbility, which is beneficial for more consistent quality control. However, powder with tightly controlled particle size ranges have been reported to result in both higher and lower powder bed packing densities, which could be caused by various factors such as powder spreading mechanisms and particle shapes [64, 65]. On the other hand, it is generally agreed that narrow particle size distribution is beneficial for mechanical properties of the parts. With the overall limited understanding with the AM powder characteristics, extra caution should be paid when investigating the powder bed property- part quality/performance relationship, since unspecified powder characteristics might potentially skew the results.

Regardless of the quality of the powder bed, there always exist particle shape and size variations, which result in intrinsic process instability with powder bed fusion AM processes. For most computational models that attempt to simulate the thermal or thermomechanical characteristics of the powder bed fusion AM processes, the powder bed properties are considered to be spatially uniform, while the effect of local property variations at individual powder particle scale was largely ignored. However, for features with small dimensions such as thin wall features, the influence of individual powder particles can become significant. Simulation based studies have shown that capillary forces exert significant effect in the formation of continuous melting pool and therefore continuous solid tracks upon solidification [66–69]. As shown in Fig. 5.26, for a single-layer single-track feature, the ignorance of capillary forces between the liquid phase and the powder bed results in significantly altered melting pool evolution history and therefore highly inaccurate results. When the dimensions of the melting pool are comparable to the powder particle sizes, the capillary force dominates the behavior of the melting pool, and the solidified track exhibit Plateau-Rayleigh instability as shown in Fig. 5.27, which results geometrical dimension in variations and increased surface roughness. In certain circumstances, under the effect of Plateau-Rayleigh instability, the liquid phases could also segregate into spherical droplets and cause the so-called balling effects [70–72]. Both experimental studies and simulations have demonstrated the rather complicated effect of process parameters on the instability. Input energy density, which can be expressed as Eq. (5.3), has been shown to exert significant

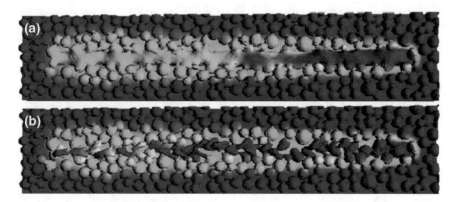

Fig. 5.26 Single-track simulation of laser melting of powder bed. **a** Capillary force considered, **b** No capillary force considered [68]

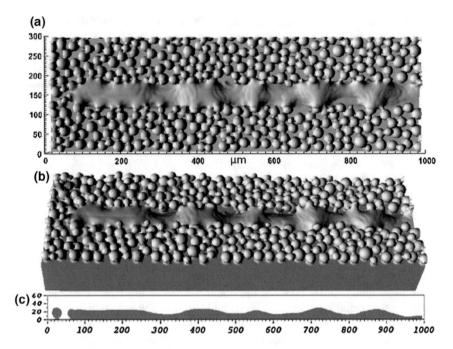

Fig. 5.27 Instability of the single-track melting [68]

effect on the balling phenomenon during the melting-solidification of the powder bed. In Eq. (5.3), E is the energy density in J/mm^3, P is the energy beam power, v is the scanning speed, t is the energy absorption thickness, d is the energy absorption width, and n is the number of repeated scanning. While excessive energy density could potentially cause balling due to the elongated existence time of the liquid phase, which leads to more sufficient coalescence driven by surface energy [73]. However, this rule can be complicated by the interaction between the liquid phase and the powder bed. When the process takes place under high energy density and very slow energy beam scanning speed, the balling phenomenon could be avoided, which is suggested to be contributed by the widened melting track width and wetting of liquid phase with the surrounding powder bed [67, 71–73]. In order to counter the process instability caused by both Plateau-Rayleigh instability and the stochastic powder bed characteristics, it is suggested that thinner layer thickness and slower beam scanning speed should be adopted when the total input energy density is held constant [67]. Such process setups could potentially result in narrower track width and improved surface finish as shown in Fig. 5.28 due to more sufficient melting and liquid continuity in the vertical directions. The significance of surface capillary effects in the powder bed melting process also helps understanding the importance of powder bed density, as higher powder bed density would facilitate better wetting of liquid and the alleviation of balling effects [74].

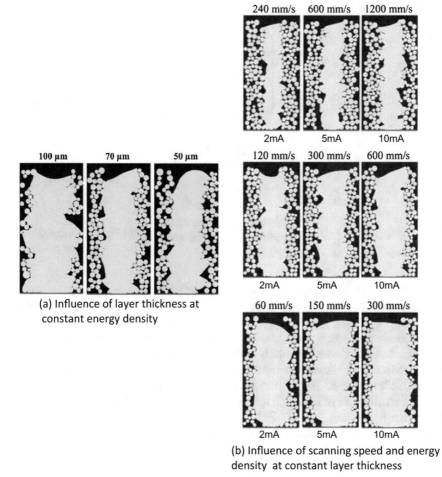

240 mm/s   600 mm/s   1200 mm/s

2mA   5mA   10mA
120 mm/s   300 mm/s   600 mm/s

2mA   5mA   10mA
60 mm/s   150 mm/s   300 mm/s

2mA   5mA   10mA

100 µm   70 µm   50 µm

(a) Influence of layer thickness at constant energy density

(b) Influence of scanning speed and energy density at constant layer thickness

Fig. 5.28 Effect of process parameters on the multi-layer single wall formation [67]

$$E = \frac{nP}{vtd} \tag{5.3}$$

For the melting pool dynamics, other important factors that are sometimes overlooked are the melting pool convection, the evaporation and the key–hole effects. Within the melting pool, due to the energy intensity distribution and rapid motion of the energy sources, the temperature distribution of the melting pools is not uniform. Due to the temperature gradient, there exist internal convective flow in the melting pool, which is termed Marangoni convection [75]. Marangoni convection affects the morphology of the melting pool and therefore the solidified track. For many

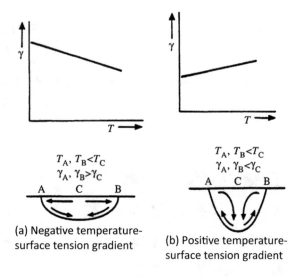

Fig. 5.29 Marangoni effect in powder bed melting pool [75]

(a) Negative temperature-surface tension gradient

(b) Positive temperature-surface tension gradient

metal alloys that have negative temperature–surface tension gradients, the occurrence of Marangoni convection would result in outward flow of the molten liquid as shown in Fig. 5.29a, since the temperature at the beam center is higher than the surrounding areas. As a result, the melting pools shape become wider and flatter. On the other hand, when materials with positive temperature–surface tension gradients, Marangoni convection results in a narrower and deeper melting pool as shown in Fig. 5.29b. Beside the effect with the melting pool characteristics, Marangoni effect plays important role in producing continuous smooth track at high input energy densities [76, 77], and it could also be intentionally utilized to create turbulent effect and to break the surface oxide layers of the substrate for reactive alloys such as aluminum [78].

Beside convective effect, various factors such as the evaporation effect [79] and keyhole effect [80, 81] could also affect the characteristics of the melting pool. For example, when excessive energy input causes the keyhole effect, the resulting melting tracks exhibit very deep penetration and relatively narrow width. In addition, when the keyhole mode forms, the process is also more prone to generate vapor cavity defects within the melting track [81]. As the melting pool morphology can be used as a direct indicator of the fabrication quality for directed energy deposition AM processes [82], the same approach has been attempted for powder bed fusion processes in various studies [83, 84]. On one hand, it has been reported that there exist significant correlations between the melting pool morphology such as length to width ratios and the microstructure of the solidified structures [84], which can be utilized in combination with the in-process melting pool monitoring to potentially realize closed-loop process control of metal AM processes. On the other hand, so far such approach has been largely restricted to the process characterization of single-layer single-track structures, and its effectiveness has not yet been verified with complex 3D structures.

Referring to the energy density equation Eq. (5.3), during the process development of a certain material, the layer thickness are usually fixed at small values to
facilitate the overall quality control as long as fabrication time is not a paramount
priority. Therefore, a common experimental-based approach for process development is to make adjustment with beam power, scanning speed and scan spacing.
When the scan spacing is fixed, various studies have shown that there exist optimal
beam power/scanning speed ratio envelopes in which the powder bed fusion AM
processes could yield best overall qualities including part densities and strength. As
shown in Fig. 5.30, the at medium power/velocity ratio ranges, the process could
generate continuous melting tracks with minimum macroscopic defects. On the
other hand, at either high or low power/velocity ratios, defects such as balling
effects and lack of fusion could occur. Note that most of such power/velocity
ratio-quality relationships were developed under particular sets of other material
and process parameters, therefore the specific values usually possess only limited
values. However, the general trends usually hold valid for most common AM metal
materials, which could help developing efficient experimental designs during the
process optimizations.

For more complex geometries, even with constant layer thickness and the same
material powder bed, the power/velocity ratio could not be used alone sufficiently as
design indicators. Instead, there could exist significant differences between the
parameter sets with same power/velocity ratios but different absolute values of
powers (or velocity, equivalently). For example, for overhanging features, at the
same energy density levels, slower scanning speed and lower power are preferred,
which help to avoid the formation of deep pool and therefore allows for more
sufficient melting pool evolution and the reduction of balling fragmentation before
solidify [69, 86]. Similarly, for the generation of thin-wall features using laser
melting processes, both lower power/low velocity and high power/high velocity
combinations could result in finer wall thickness resolution [87], although they
might be driven by slightly different mechanisms. Even between similar types of

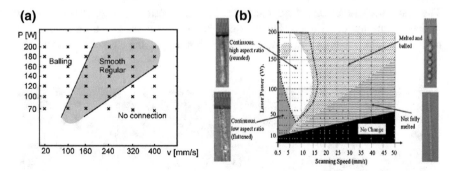

Fig. 5.30 Typical process map for laser melting processes. **a** Fe-Ni-Cu-Fe$_3$P alloy [72]. **b** 316L
stainless steel [85]

features such as thin wall features and thin beam features, the optimal power/velocity combinations for best resolutions are different [88].

Some of the other process parameters, such as scanning line spacing, layer thickness and protection atmosphere, usually have relatively straightforward effects on the process qualities. For example, larger scanning line spacing and layer thickness usually corresponds to higher tendency of porosity generation and therefore reduced qualities [89]. Figure 5.31 illustrates the effect of increasing scanning line spacing on part qualities. The type of atmospheric gas during the laser melting process could also potentially affect the part densities, with argon gas corresponds to slightly higher part density under the same energy input densities compared to nitrogen [89]. This might be contributed by the lower thermal conductivity of argon and therefore less heat loss during the fabrication.

One of the most interesting issues in the process design for powder bed fusion AM is the optimization of the scanning strategies. Due to the high energy density of the laser and electron beams, the beam scanning sequence and patterns essentially determine the distribution of thermal gradients and temperature fields. Theoretically, an ideal scanning strategy should realize the simultaneous heating of the entire processed area. However, due to the non-uniform heat transfer conditions, such as the differential thermal dissipation rate at the boundaries versus the center of the parts, it is impossible to archive uniform temperature field with uniform heating. Most current powder bed fusion systems address such issue by using multiple scanning themes in combination. One common scenario is to use contour scanning and hatch scanning in combination, since the thermal transfer conditions between the boundary and the interior areas are usually most significant. Figure 5.32 shows the real-time thermal imaging of the electron beam melting (EBM) process during both the contour and hatch scanning, and the differences of temperature fields are obvious. In this particular case, the contour scanning adopted a "multi-spot" strategy, which generates contour heating at multiple locations essentially simultaneously from the perspective of the thermal diffusion time. With the rapid scanning capability of electron beam, both the number of the heating locations and the advancing speed of these individual locations can be controlled, which is also illustrated in Fig. 5.32. The resulting temperature field therefore exhibits thermal gradients between the boundary and the other areas. On the other hand, with hatch scanning a large and more continuous melting pool forms in the interior of the processed areas, and

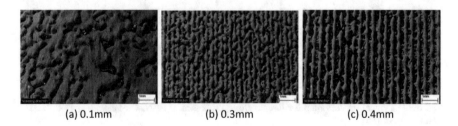

(a) 0.1mm   (b) 0.3mm   (c) 0.4mm

Fig. 5.31 Effect of scanning line spacing on part quality [89]

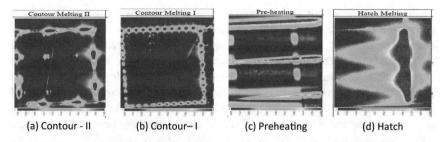

| Contour Melting II | Contour Melting I | Pre-heating | Hatch Melting |

| (a) Contour - II | (b) Contour– I | (c) Preheating | (d) Hatch |

Fig. 5.32 Different scanning themes in EBM process [90]

therefore the thermal gradients largely occur at the interior areas that evolves as the melting pool advances. With laser melting processes, due to the relatively slow scanning speed, currently it is infeasible to realize "multi-spot" melting pool control, therefore, the temperature fields in laser melting exhibit different distributions, and the processes usually exhibit higher overall thermal gradients.

As shown in Fig. 5.33, for a simple one-direction scanning pattern, the energy beam quickly scans along the primary direction before progressing by one scan spacing index and moving to the next scan line. Under this scanning strategy, the thermal gradient are primarily contributed by the inter-scan line gradient, which is influenced by the time delay between the two consecutive heating events at any certain location, and the in-line gradient as a result of the scanning motion of the energy source. The in-line gradients are usually less significant in comparison since the gradients transition is more gradual. On the other hand, the inter-scan line gradient could be rather steep as the time delay (t_1 and t_2 as shown in Fig. 5.33) becomes large.

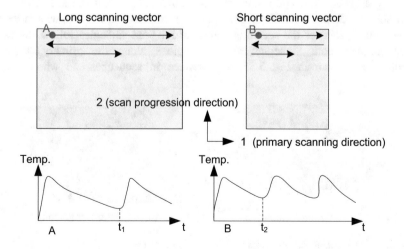

Fig. 5.33 Relationship between thermal gradient and the scanning vector length

Two approaches with the scanning vectors are regularly employed in order to reduce the overall thermal residual stress of the structures, which are the optimization of scanning vector directions and the reduction of scanning vector length. For the scanning vector direction manipulation, alternating scanning strategies and angular offset scanning strategies are most commonly employed by commercial powder bed fusion systems in order to reduce process anisotropy and the overall thermal distortion. As shown in Fig. 5.34, with alternating scanning strategy, the scanning vector alters between x-direction and y-direction between consecutive layers, therefore alleviating the thermal residual stress accumulation in a particular planar location. It was reported that alternating scanning strategy could result in as much as 30% reduction of the thermal distortion compared to single-directional scanning [91]. With angular offset scanning strategy, the scanning vector rotates by a fixed angle in a predetermined direction for each consecutive layer, which facilitates the generation of more "isotropic" thermal residual stress distribution patterns for simple geometries.

Shorter scanning vector length in the primary scanning direction usually corresponds to shorter interval between heating cycles for a particular location and therefore smaller thermal gradient. However, the scanning vector could not be indefinitely shortened, and for complex geometries it is often impractical to optimize for scanning vector due to the other process and design constraints. Another more feasible alternative is to split the scanning region into smaller sub-regions with shorter overall vector lengths. Although in concept this approach is promising, due to the added complexity with the heating pattern and the consequent temperature field evolution, it becomes more difficult to design and analyze the process.

A limited number of experimental-based investigations had been performed on this issue. As shown in Fig. 5.35, several scanning strategies that combine both vector direction offset and region-dividing methods were investigated along with the single-direction scanning strategies (x-scanning, strategy 1 and y-scanning, strategy 2) as benchmark for their effects on thermal residual stress. In strategy 3 and strategy 5, the scanning region are divided into sub-regions with different dimensions, however the scanning sequence for both strategies follow the same scenario. For both strategies, the sub-regions along the primary direction (x-direction shown in Fig. 5.35) are scanned in sequence, and when all the

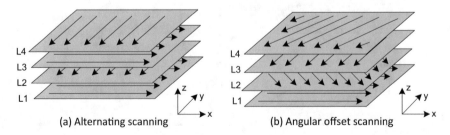

(a) Alternating scanning   (b) Angular offset scanning

Fig. 5.34 Typical scanning strategies with laser melting systems

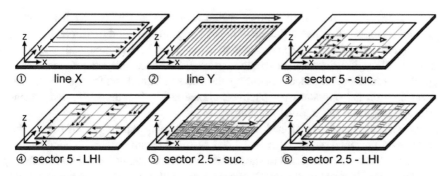

Fig. 5.35 Different scanning strategies in laser melting process [71, 72]

sub-regions along that direction are scanned, the scanning progresses by one sub-region dimension in the secondary direction (y-direction) and then follow the same scanning sequence. In addition, for each consecutive sub-region, the scanning vector rotates by 90°. On the other hand, in strategy 4 and 6, there also exists difference between the sub-region dimensions, but both strategies follow a random pattern in selecting the sub-regions to scan. Also, for both strategies, the scanning vector rotation of 90° is applied between consecutive sub-regions to be scanned.

The effect of different scanning strategies on the thermal residual stress induced distortions is shown in Fig. 5.36 for average distortions in both x and y directions. It could be clearly seen that the single-directional scanning strategies result in significantly higher thermal distortions in the directions perpendicular to the primary scanning vector directions. Similar trend is also obvious for the sub-region sweep scanning strategies (i.e. strategies 3 and 5), as there still exist a defined overall scanning progression direction (x-direction). However, by introducing angular offset with the scanning vector and smaller vector length, the magnitude of the thermal distortion is significantly reduced. On the other hand, the random sub-region scanning strategies (i.e. strategies 4 and 6) result in more uniform

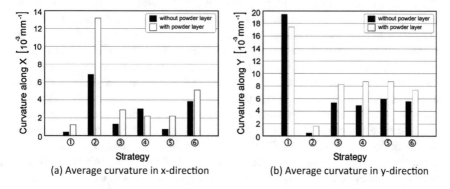

Fig. 5.36 Effect of different scanning strategies on thermal distortions [71, 72]

thermal distortion in both directions, which is partly contributed by the random scanning. On the other hand, the effect of reducing sub-region dimensions becomes less significant when the dimensions become smaller. For practical purposes, smaller sub-region dimensions would also result in more sub-regions and consequently more complex process planning and control. Therefore, from both thermal stress control and operation perspectives, choosing a moderate sub-region dimension appears to be the optimal practice.

As was previously mentioned, size dependency plays an important role in the design and characterization of powder bed fusion AM structures. Smaller feature dimensions usually correspond to shorter scanning vector length and therefore favor smaller overall thermal gradients. On the other hand, the end-of-vector effect caused by kinetic inertia of the laser scanning hardware as well as the altered heat dissipation conditions could result in local or global fluctuation of thermal gradients. Such effects have been reported in multiple works for laser sintering but has not been systematically studied for metal melting processes. Size dependency of material properties for bulky metal structures made by powder bed fusion AM is currently unknown, however, for thin features such as thin walls and thin beams, the microstructure of the structures exhibit significant size dependency. As shown in Fig. 5.37, under the same process parameters, the columnar grain widths for thin beam structures with varying dimensions and build orientations exhibit non-negligible variations. In addition, there also exhibits grain size variations between the exterior (surface) and the interior of a beam. Such microstructural variations likely correspond to mechanical property variations, which still remain largely unexplored.

In the design of mechanical properties for powder bed fusion processes, there exist an abundant of literatures that are mostly based on experimental studies. Anisotropy and quasi-static properties are among the most commonly characterized issues, although there exist some discrepancies with the conclusions. For both laser melting and electron beam melting based powder bed fusion processes, extensive literatures exist in developing "optimal" process parameters for relatively standardized structures such as ASTM tensile coupons for various materials, which could often be used as baseline properties in the evaluation of process parameters for an arbitrary geometry. Table 5.5 summarizes some of the reference mechanical properties of multiple common AM metal materials using either electron beam melting or laser melting based on previous experimental studies. Note that the values listed in Table 5.5 represent as-fabricated properties, therefore caution must be taken when these reported values are used for comparison since in order to minimize thermal residual stress parts fabricated by laser melting processes are often subjected to stress relief heat treatments in practice, which could result in mild increase of elongation. Comparing the values from Table 5.5 with the manufacturer's material specifications (Table 5.1), the values are generally in agreement since in many of these experimental studies standard tensile testing methods were adopted. However, there also exist some discrepancies, which was likely a result of the further fine tuning of the process parameters by individual researchers.

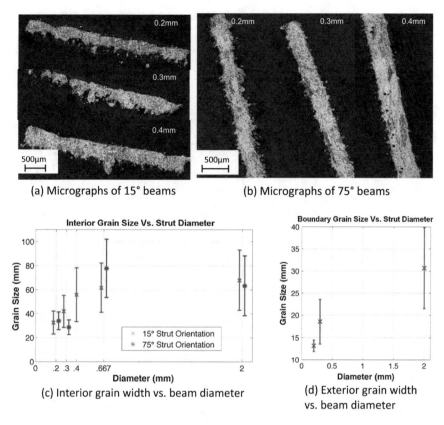

Fig. 5.37 Size dependency of microstructure for thin features [92]

In addition, depending on the sample orientations during fabrication, the mechanical properties of these materials could also differ from the listed values.

For metal powder bed fusion processes, anisotropic mechanical properties are reported by most literatures, which is a direct result of the anisotropic microstructure generated during the process. For both electron beam and laser based processes, columnar grains form along the build direction (Z-direction) in most cases unless the scanning strategies are intentionally designed to alter the dominant thermal gradient directions. In most studies, it was reported that the horizontally fabricated tensile coupons (i.e., XY or YX orientation per ASTM F2971-13) exhibit higher quasi-static mechanical properties (e.g., yield strength, ultimate strength, elongation) compared to vertically fabricated samples (i.e., Z-direction). However, several studies also reported opposite conclusions with the horizontal fabrication being the weakest orientation, although it was unclear whether this was partly contributed by certain un-specified process setting.

Table 5.5 Benchmark mechanical properties of AM metals from experimental studies

Material	E-modulus (GPa)	Yields strength (MPa)	Ultimate strength (MPa)	Elongation (%)
Ti6Al4V-SLM	109.9 [93] 109.2 [94] 94 [95]	736 [93] 1110 [94] 1125 [95]	1051 [93] 1267 [94] 1250 [95]	11.9 [93] 7.3 [94] 6 [95]
Ti6Al4V-EBM	118 [96] 104 [43]	830 [96] 844 [43]	915 [96] 917 [43]	13 [96] 8.8 [43]
CP-Ti-SLM	80 [97]	555 [97]	757 [97]	19.5 [97]
AlSi10Mg-SLM	73 [98] 68 [99]	243 [98]	330 [98] 391 [99]	6.2 [98] 5.2 [99]
Ni-263-SLM	163 [100]	818 [100]	1085 [100]	24 [100]
IN718-SLM	204 [102]	830 [101] 898 [102]	1120 [101] 1141 [102]	25 [101] 22.6 [102]
IN625-SLM	204 [104]	571 [103] 800 [104]	915 [103 1030 [104]	49 [103] 10 [104]
IN625-EBM		410 [105]	750 [105]	44 [105]
316L-SLM	183 [106]	465 [106]	555 [106]	13.5 [106]
CoCrMo-EBM		510 [108] 635 [110]	1450 [108] 754 [110]	3.6 [108] 3.1 [110]
CoCrMo-SLM		503 [107] 749 [108] 884 [109]	951 [107] 1061 [108] 1307 [109]	15.5 [107] 4.8 [108] 10.2 [109]
Ti-48Al-2Cr-2Nb (γ-TiAl)-EBM	175 [111]	360 [111]	480 [111]	1.2 [111]
17-4-SLM		600 [112] 570 [113]	1300 [112] 944 [113]	25 [112] 50 [113]

Fatigue properties for the as-fabricated powder bed fusion metal materials are generally significantly lower compared to the wrought ones, which is largely contributed by the combination of internal defects, surface roughness and thermal residual stress. For example, Figs. 5.38 and 5.39 show some of the compiled results for Ti6Al4V from recent experimental studies for both laser and electron beam based systems. As-received AM Ti6Al4V samples exhibit about 0.1–0.15 σ_y for endurance level with σ_y representing the quasi-static yield strength of the material, which is only 25–38% of the typical endurance level for solid wroughtTi6Al4V (0.4 σ_y). It was suggested that for bulky structures the internal defects become dominant factors for the deterioration of the fatigue performance of these AM parts, while thermal residual stress issues can be largely alleviated via proper heat treatment. On the other hand, when process parameters and post-processes are properly designed, surface defects caused by surface sintering and process instability becomes a significant contributor for crack initiation and the premature part failure under cyclic loading. Such issue could become more significant when

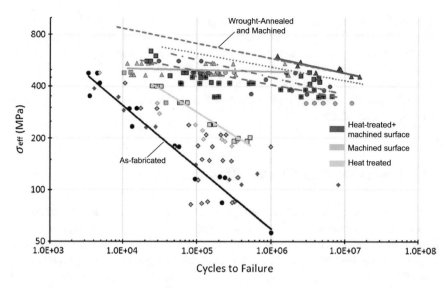

Fig. 5.38 Uniaxial fatigue performance of laser melted Ti6Al4V parts by laser melting at $R \approx 0$ from various experimental studies [118]

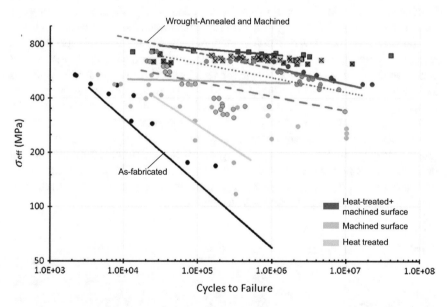

Fig. 5.39 Uniaxial fatigue performance of laser melted Ti6Al4V parts by electron beam melting at $R \approx 0$ from various experimental studies [118]

lightweight structure are designed, which generally contain large specific surface areas [114–116]. Due to the difficulty of surface treatment with AM structures, surface defects appears to be a more difficult challenge for the fatigue performance improvement of AM metal parts. In addition, the columnar grain morphology of both the laser and electron beam melted structures could potentially favor low cycle fatigue resistivity and contribute to the surface crack dominated failure behavior under high cycle fatigue [117].

5.6 Design for Lightweight Structures

One of the objectives of structural design is to minimize the mass consumption and maximize the utilization efficiency of the materials. Therefore, lightweight structure design has always been sought after for almost all the engineering designs. Lightweighting brings about various technical advantages such as high strength to weight ratio, high energy absorption per weight ratio, low thermal conductivity, and large surface area to volume/weight ratio. These attributes could in turn translate into various economical and environmental benefits such as product reliability, system energy efficiency and product sustainability. However, as lightweight designs often involve high level of geometrical complexity, the realization of these designs has been a challenging task with traditional manufacturing technologies. It has been widely recognized that AM technologies possess unique capabilities in realizing lightweight designs with little penalty from geometrical complexity, and extensive demonstrations are available for various applications such as fashion, arts and biomedicine as shown in Fig. 5.40 [119–123]. However, the design of these lightweight structures beyond esthetic purposes is in general not well understood by most designers. The lack of understanding on the relationship between various engineering performance requirements (e.g. mechanical properties, thermal properties, biological properties, etc.) and the geometrical design often prevents efficient design of lightweight structures for functional purposes. In addition, there also exist little literature in the guidance of optimal process selections in the fabrication of these structures utilizing different AM technologies.

(a)   **(b)**   **(c)**   **(d)**

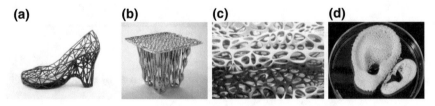

Fig. 5.40 Lightweight designs realized via AM in art, fashion and biomedicine. **a** Fashion shoes [124]. **b** Table [125]. **c** Skin shell [126]. **d** Artificial ear [127]

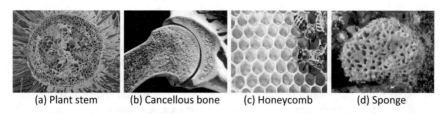

(a) Plant stem   (b) Cancellous bone   (c) Honeycomb   (d) Sponge

Fig. 5.41   Natural lightweight structure designs

Nature has been providing countless examples of lightweight designs that serve different purposes. As shown in Fig. 5.41, structures such as plant stem and bone tissues not only serve load-bearing purposes but also function as conduit for biological fluids. In addition, these structures often exhibit local optimization. For example, in typical bone tissues the bone porosities towards exterior surface tend to be lower in order to provide higher torsional and bending stiffness. For another example, in the plant stem structures the sizes of the pores at the centers are significantly larger compared to the ones near the skins, which correspond to lower mechanical strength but better fluid flowability that benefits nutrient transportation through the plants. Similarly, the topology of the honeycomb structure also serves multiple functionality including high stiffness and large pore volumes, which makes it ideal as both a stronghold and a storage for honeybee colonies.

Most man-made lightweight structures have considerably less sophisticated designs. In many of these designs, the functionality of the structures is often simplified to allow for relatively simple lightweight geometries to be used. For example, in the design of space filling stiffener for secondary structures in aircrafts such as wall panels, sandwich panels with regular honeycomb cores are often utilized due to their high uniaxial stiffness-to-weight ratio. Therefore, the design of the honeycomb cores could be largely simplified into the single-objective design of wall thickness or honeycomb unit cell width once the material is determined. This allows for a pure experimental-based design approach that is often adopted by designers. However, in order to achieve more optimized designs for multi-functionality such as the ones shown in Fig. 5.41, a more sophisticated design theory must be employed.

One of the objectives of AM lightweight design is to achieve the design for functionality (DFF), which implies that the design should be driven by engineering specifications instead of manufacturability limitations. The underlying prerequisite for such approach is that the relationships between the engineering specifications and the upstream design components are well quantified, which in turn include both the geometry design and the process design.

While there exist various geometry design and optimization approaches and tools that generally allows for the creation of design models with improved lightweight performance, none of them addresses the process design adequately. Most of the lightweight design tools treat material as an ideal isotropic material and focus on geometry optimization only, which results in a significant design deviation from reality. As previously mentioned, one of the unique characteristics of AM is that the material properties are often process and geometry dependent. Such compound effect has significant impact in the design practice of lightweight structures, since these structures often have geometrical features that have varying dimensions and therefore potentially varying material properties. In addition, due to the intrinsic quality variations with many AM processes with small-dimension geometries, the difficulty of achieving accurate design is further signified. The process design for lightweight structures is an ongoing research area in AM community, and some of the preliminary results are presented in this book for the reader's reference.

5.6.1 Geometric Design for Lightweight Structures

In the design of AM lightweight structures, two basic approaches are most commonly employed currently, which are topology optimization based design and cellular structure design. Each approach possess certain advantage and disadvantage compared to each other.

5.6.1.1 Topology Optimization

Although it was suggested that the concept of topology optimization was first introduced as early as 1901, the method did not receive significant attention until mid-1980s [128]. In a nutshell, topology optimization is a mathematical tool that focus on achieving an optimized material distribution within a given design space with the objective of maximizing/minimizing certain constraint criteria such as weight, stiffness, compliance and conductivity. The implementation of topology optimization is often realized through finite element methods, since both are based on the discretization of design space and voxel based analysis. Unlike sizing optimization and shape optimization as shown in Fig. 5.42, topology optimization does not prescribe design topology, and is therefore capable of yielding optimized geometrical solutions [129]. An example of topology optimization is shown in Fig. 5.43, in which a minimum compliance/deflection optimization was performed on an overhanging beam with one fully constrained end [130]. During the optimization, individual voxels were selectively removed by the algorithm, and the resulting structure would be checked against the target objective to evaluate whether any improvement was made. This iterative optimization process continues until either a satisfactory result is achieved or that the improvement diminishes below the predefined threshold.

Various topology optimization algorithms have been reported in the design of AM structures such as solid isotropic microstructure with penalization (SIMP), bidirectional evolutionary structural optimization (BESO) and proportional topology optimization (PTO) [128, 131–134]. Some methods such as BESO perform

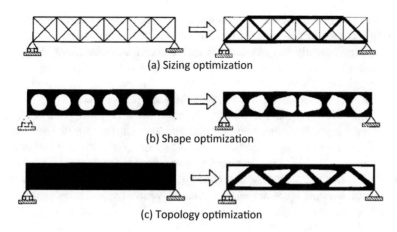

(a) Sizing optimization

(b) Shape optimization

(c) Topology optimization

Fig. 5.42 Comparison of different optimization algorithms [129]

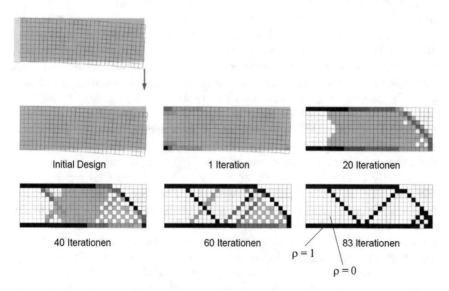

Fig. 5.43 An example of a 2D topology optimization problem [130]

optimization by adding or removing fully dense voxels, which means that for each voxels in the design space there are two possible status, which is either "1"—fully filled with material, or "2"—void with no material. On the other hand, other methods such as SIMP and PTO optimize the structures with variable-density voxels (which reduces to elements in 2D cases), which means that a relative density (i.e. density of the voxel divided by the density of the solid material, ρ/ρ_s) between [0,1] could be considered during the optimization, and the mechanical properties of these variable-density voxels are predefined by either linear mixing rule or other physically reasonable rules. It was suggested that improved global optimal solutions could be achieved by allowing for variable-density voxels in the designs [133, 135]. However, traditionally the physical interpretation of variable-density voxels was rather difficult since it could not be realized via most manufacturing processes methodologically. Even with AM technologies, currently the systems that could realize variable-density designs are limited. Material jetting systems (e.g., Stratasys PolyJet and 3D Systems MultiJet) with multi-materials that possess different material properties (e.g. one soft material and one hard material) could closely imitate continuous density/property variations by mixing the two base materials into "digital" materials [136, 137]. Figure 5.44 demonstrates the difference between the two optimization approaches with a classical overhanging beam problem with the objective of minimizing strain energy (SE), in which the variable-density is represented by the gray scale of the elements [133]. Note that the generation of the partially dense voxels could be completely subdued in the SIMP method by introducing adequate penalty [133].

Since currently most metal structural components are fabricated via powder bed fusion AM, one possible alternative to realize variable-density design is to introduce porosities in the process. This could be realized through the change of process processes [138] or intentionally introduce porosities in the design models [139, 140]. Using the porosity control approach, the porosity–property relationship could be predicted through classical cellular theories, which will be introduced in more details in the following section. Such attempt that combines topology optimization and variable-density cellular structure have been demonstrated as shown in Fig. 5.45 [141, 142]. Although preliminary, such approach has the potential of effectively combining the advantages of both topology optimization and cellular design approaches as will be elucidated in the following discussions.

Currently there exist a number of software package that offer topology optimization such as Altair Optistruct, Altair solidThinking, ANSYS Genesis and Autodesk Shape Generator. Most software are relatively friendly to users who are used to finite element software environment. The design space as well as the boundary conditions (e.g. loading, constraint, temperature, etc.) must be defined. The most common optimization objectives include the minimization of weight subject to the constraint of minimum safety factor or stiffness and the maximization of stiffness or compliance subject to the constraint of total mass. Most of the

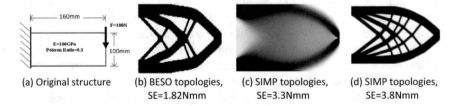

(a) Original structure   (b) BESO topologies, SE=1.82Nmm   (c) SIMP topologies, SE=3.3Nmm   (d) SIMP topologies, SE=3.8Nmm

Fig. 5.44 Comparison of BESO and SIMP method [133]

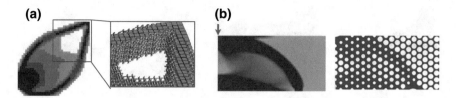

Fig. 5.45 Examples of utilizing cellular structures in topology optimization for variable densities. **a** 3D cubic lattice [141]. **b** 2D hexagon [142]

optimization solvers could handle static mechanical problems satisfactorily without issues such as checkboard pattern, however few currently provide little or limited capabilities with multi-physics problems and some dynamic loading problems such as fatigue problem, which are subjects of intensive research development [143, 144].

Another challenge with the current topology optimization is the manufacturability issue. Even though some of the topology optimization software provide certain capabilities for manufacturability constraints such as constant cross section for extrusion features, for AM systems no such capability currently exists. For example, for the bracket part shown in Fig. 5.46, the topology optimized geometry may contain features with very small dimensions that are not realizable with the metal powder bed fusion AM processes, or features with drastic cross-sectional dimension change that potentially causes thermal stress concentration during the fabrication processes. Even though some software do offer additional design constraint of minimum dimension, the manufacturability issues caused by the lack of overall shape definition could not be fully offset. In addition, most of the topology optimization software still lacks the ability to handle anisotropic materials and materials with geometry-dependent material properties. Again such issue could be partially overcome by introducing cellular design into topology optimized structures, which introduces certain level of geometrical definition that helps to predict the manufacturability.

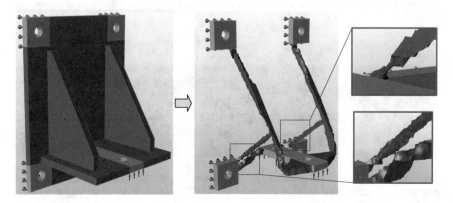

Fig. 5.46 Potentially non-manufacturable features generated by topology optimization

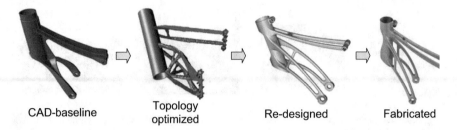

CAD-baseline   Topology
                      optimized   Re-designed   Fabricated

Fig. 5.47 Topology optimization as design reference [145]

Currently topology optimization method is most commonly used to provide a reference topology for structural optimization. As shown in Fig. 5.47, topology optimization will be performed on the baseline design of a bicycle seat post, which creates a non-manufacturable topology optimized design. Additional manual designs and verifications would then follow to finalize the component designs. Such approach serves to improve the current designs, but also imposes significant constraint to the designers since the redesign from topology optimized references is often a less straightforward practice. On the positive side, it could be reasonably expected that with the rapid development of computational software such manual design iterations could be automated in the near future.

5.6.1.2   Cellular Structure Design

Cellular structures are generally defined as networks of interconnected solid beams or walls with included voids. Classical theory for cellular structures has been well developed by various groups [146, 147], which focuses on both stochastic cellular structures and honeycomb cellular structures.

Stochastic cellular structures are often referred to as foams, and they generally exhibit random porosities and pore characteristics including pore size and pore shapes. Figure 5.48 shows some of the commercially available metal stochastic

(a)   (b)   (c)   (d)

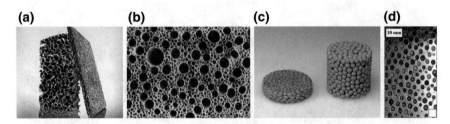

Fig. 5.48 Stochastic cellular structures. **a** Cymat foam by gas melt injection [152]. **b** Solid-gas eutectic solidification foam [150]. **c** Sintered hollow spheres [150]. **d** Al-SS casted foam [153]

cellular structures, which are mostly made by traditional processes such as casting, sintering and foaming [148–151]. Due to the randomness of geometries, relative density is used as the primary design parameter.

Relative density is defined based on the volume ratio between the cellular structure and its geometrical bounding volume, which is the total volume of solid and void combined. Therefore, the relatively density ρ_r and the porosity η always satisfy the following relationship.

$$\rho_r + \eta = 1 \tag{5.4}$$

For cellular components, the geometrical bounding volume could be easily identified as the overall bounding spaces of the components. However, for cellular structures that are constructed by unit cells, the determination of geometrical bounding volume is closely related to the spatial pattern of these unit cell geometries. For example, for the unit cell shown in Fig. 5.49a, the two geometrical bounding volumes shown in Fig. 5.49b, c would correspond to two different cellular structures.

For stochastic cellular structure, based on whether the pores are interconnected, there exist two types of porosities. Open cell structures have interconnected pores (Fig. 5.48a), while closed cell structures have isolated pores (Fig. 5.48d). In general open cell structures exhibit lower mechanical strength compared to closed cell

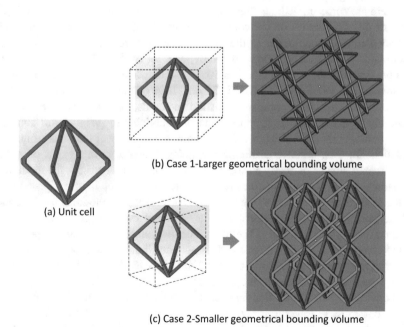

(a) Unit cell

(b) Case 1-Larger geometrical bounding volume

(c) Case 2-Smaller geometrical bounding volume

Fig. 5.49 Relationship of geometrical bounding volume and cellular pattern

structures. However, in certain applications such as biomedical implants and cat-alyst, the pore interconnectivity becomes beneficial.

In general the mechanical properties of the open cell cellular structures could be predicted by their relative densities reasonably well by following the Gibson–Ashby theory [146]:

$$\frac{E}{E_S} = C_1 \rho_r^2 \tag{5.5}$$

$$\frac{\sigma}{\sigma_S} = C_2 \rho_r^{1.5} \tag{5.6}$$

$$\frac{G}{G_S} = C_3 \rho_r^2 \tag{5.7}$$

where E, E_S, σ, σ_S, G, G_S are the elastic modulus, yield strength and shear modulus of the cellular structure and the solid material, respectively, and C_1, C_2, and C_3 are constants that varies between different cellular structures which are usually deter-mined through experimentation. For stochastic cellular structures made by Al alloys, C_1, C_2 and C_3 are approximately 1, 0.3 and 0.4 respectively. Although in some literatures modulus and strength equations with slightly different exponential values were reported, Eqs. (5.5–5.7) generally hold valid for all cellular structures even including non-stochastic ones.

There are some immediate conclusions that could be made from Eqs. (5.5–5.7). First of all, the moduli of the cellular structure is more sensitive to the relative density variations compared to the yield strength, therefore could be potentially controlled within a larger range. Secondly, as a lot of the cellular structures have relative densities that are smaller than 50%, their elastic modulus, yield strength and shear modulus would fall below 25, 35 and 10% of the solid material properties respectively. Therefore, compared to solid materials, cellular structures are con-siderably weaker. Thirdly, it appears that the modulus to weight ratio and yield strength to weight ratio of the cellular structure are both lower than that of the solid materials. In fact, when considering the uniaxial elastic compression of a 2.5D cellular extrusion of perfectly elastic material as shown in Fig. 5.50, the modulus of the structure is proportional to the cross-sectional area, which in turn is proportional to the total relative density of this structure. As a result, the modulus forms a linear relationship with relative density and the structure consequently exhibits higher modulus to weight ratio. For 3D cellular structures, since the porosity perpendicular to the compressive loading direction does not contribute to the modulus/stiffness of the structure, the overall modulus to weight ratio of the cellular structures will always be lower compared to the 2D ones.

It could be easily shown that it is generally the case that the mechanical property to weight ratio is lower for 3D cellular structures compared to 2.5D extruded cellular structures under simple compression/tension loading cases. Such seemingly counterintuitive observation possesses critical design implications. The design for

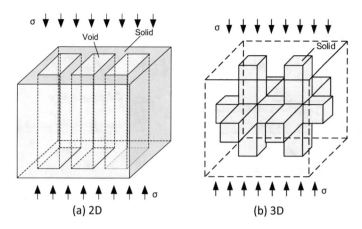

Fig. 5.50 Comparison between 2D and 3D cellular structures under uniaxial compression

lightweight structures with improved performance to weight ratio is only practical in the case where local stress variations exist. When different locations of the structure is subject to different stress levels such as in the case of a beam subject to bending, the use of continuous materials would inevitably lead to overdesign with the regions that undergo relatively lower stress levels. In comparison, the structure would become more efficient if the materials could be selectively removed in the center region of the beam, which result in either a sandwich beam with cellular core or a tube with completely hollow interior. These design examples are well-known and intuitive to design engineers, but help to demonstrate the underlying design philosophy of cellular structures.

On the other hand, uniaxial compression/tension testing is often used as an efficient method to characterize the mechanical properties of cellular structures. Unlike solid materials, cellular structures generally exhibit unique stress–strain curves under compressive or tensile loading. For a stochastic cellular structure made of elastic perfectly plastic material, the typical stress–strain curves under compressive or tensile loading are shown in Fig. 5.51 [146]. Under tensile stress, cellular structures exhibits linear elasticity initially, which is dominated by the elastic deformation within the cellular solids. Upon further deformation, some struts/walls in the cellular structures will start to yield, which consequently result in realignment of these solid components along the loading direction. Therefore, upon the initiation of yield the cellular structure will exhibit strain-hardening effect as shown in Fig. 5.51b. When the cellular fracture propagates through the cross sections, the cellular structures will fail by disintegration. On the other hand, under compressive stress, although the elastic behavior of the cellular structures still remain the same, the yield of the structure progresses much differently. Under compression individual yield struts/walls will undergo progressive deformation until they could no longer carry loads effectively. However, the failed debris

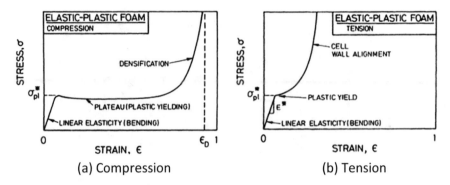

(a) Compression   (b) Tension

Fig. 5.51 Stress-strain characteristics of stochastic cellular structures under different uniaxial loading [146]

struts/walls will be trapped and compacted in the main structures, therefore continue to contribute to the overall strength of the structure. As a result, the cellular structures exhibit a plateau stage in which the stress remains largely constant while the strain accumulates as shown in Fig. 5.51a. Towards the end of the plateau stage, most of the cellular struts/walls will have failed, and the compacted structures with debris will begin to behave similar to the solid material in the so-called densification stage. The existence of plateau stage is of great interest to many energy absorbing applications with cellular structures, since it would allow for energy absorption that is almost proportional to the amount of deformation introduced into the structures.

Figure 5.52 further demonstrates the progression of compressive failure within a stochastic cellular structures. Due to the random characteristics of the pores, the crack/failure could initiate at a random location with the cellular structure. Once formed, the failure site will cause stress concentration in the neighboring regions, which become the sites for the growing failure front. As a result of the progressive failure, a failure band usually forms for stochastic cellular structures as shown in Fig. 5.52. With further compression the failure band will expand in width and eventually propagates through the entire structures.

One interesting characteristics of stochastic cellular structures is that they could potentially exhibit isotropic properties, which makes them attractive in certain applications. However, such characteristic is based on the prerequisite that the numbers of cells are large in the cellular structures so that the local property variations are averaged. In order to adequately account for such effect, size effect

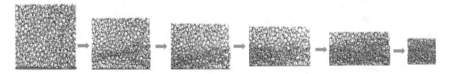

Fig. 5.52 Progression of compressive failure in stochastic cellular structure [154]

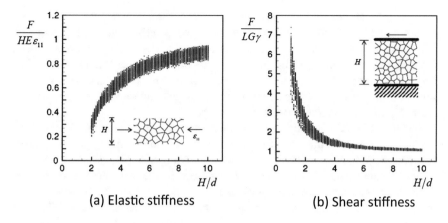

Fig. 5.53 Size effect for 2D Voronoi cellular structure [155]

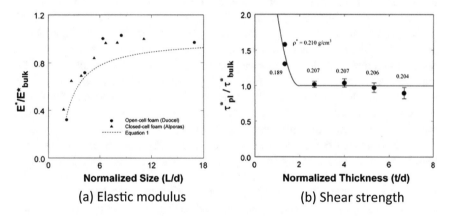

Fig. 5.54 Size effect for 3D stochastic cellular structures [156]

needs to be considered. In practice size effects also takes other factors into account such as the boundary effects. For example, when a cellular structure is constrained at the boundaries by solid skins such is the case for sandwich panels, the cellular struts/walls at the boundary will be subject to stress concentration, which could significantly change the mechanical properties of the overall structures. Another example is the free surface boundary, which does not provide any balancing force to the materials at the boundaries. As a result, these boundary struts/walls are more easily deformable and therefore could potentially offset the mechanical properties significantly. Experimental-based studies are often carried out to characterize the overall size effect. Figures 5.53 and 5.54 demonstrate the influence of size effect on elastic modulus and shear properties with 2D Voronoi cellular structures and 3D stochastic cellular structures [155, 156]. In general, with increasing number of cells the modulus and compressive strength of the cellular structure increases, while their shear strength and shear modulus decreases. The size effect tend to diminish with sufficient number of cells. Such threshold number varies among different cellular structures, however as a rule of thumb, a minimum of 8-10 cells along the loading directions are needed in order to minimize the normal stress size effects, and a minimum of 3–4 cells perpendicular to the shear loading directions are needed in order to minimize the shear stress size effects.

Compared to cellular structures with more defined topologies, stochastic cellular structures often lack sufficient design controllability, which makes accurate property design somewhat difficult. In addition, although it is possible to achieve stochastic cellular part fabrication via AM processes, the mechanical properties of these cellular structures are often inadequate for structural applications when altered process is used to produce them. Therefore, the following discussions will focus on non-stochastic cellular structures with predefined geometries. More details regarding the design of stochastic cellular structures are discussed by numerous literatures. Interested readers could refer to [150, 151] for more information.

As previously illustrated, the design of non-stochastic cellular structures could be categorized into 2.5D design and 3D design. Considering manufacturability, most non-stochastic cellular designs are open cell structures. One of the most commonly employed design approach is the unit cell based design. With this approach, a representative unit cell geometry is designed for the cellular structure and consequently patterned seamlessly in the design space to create the overall structures. It is further assumed that the boundary conditions between individual unit cells are consistent throughout the structure, which allows for the property design for one unit cell to be applied to the entire structure. Such assumption is reasonably efficient for a relatively large cellular structure with sufficient numbers of unit cells for each loading axis, since the size effect is minimized in these cases. However, there exist exceptions where the size effect does not diminish regardless of the unit cell numbers, which will be introduced in more details later. In short, such simplification treatment allows for efficient analytical modeling of non-stochastic cellular structures via unit cell analysis, but also introduces various potential modeling errors that must be addressed as well.

The design for 2.5D cellular structures is relatively straightforward. For a 2.5D extruded cellular structure, there exists a plane of unit cell pattern and a third principal direction that is out-of-plane as shown in Fig. 5.55a. The design of the unit cell geometry is required to be 2D space-filling and periodic. Such pattern design is well studied in the research of plane symmetry group. In addition, the third axis (z-axis in Fig. 5.55a) is usually designed to be perpendicular to the pattern plane so that Cartesian coordinate system could be established for stress analysis. In general the in-plane mechanical properties (i.e. mechanical properties along the direction in x-y plane shown in Fig. 5.55a) are considerably lower than the out-of-plane mechanical properties (i.e. mechanical properties along the z-direction). This could be further demonstrated by the following equations for hexagonal honeycomb structures following the axis convention used in Fig. 5.55a:

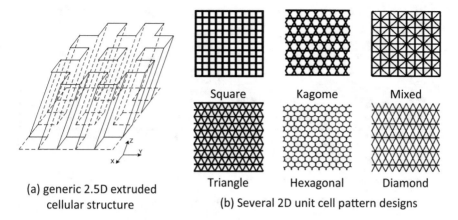

(a) generic 2.5D extruded cellular structure

Square Kagome Mixed

Triangle Hexagonal Diamond

(b) Several 2D unit cell pattern designs

Fig. 5.55 2.5D extruded cellular structure design

$$\frac{E_{x(y)}}{E_S} \propto \rho_r^3 \tag{5.8}$$

$$\frac{E_z}{E_S} \propto \rho_r \tag{5.9}$$

$$\frac{\sigma_{x(y)}}{\sigma_S} \propto \rho_r^2 \tag{5.10}$$

$$\frac{\sigma_z}{\sigma_S} \propto \rho_r \tag{5.11}$$

Comparing Eqs. (5.8) and (5.10) with Eqs. (5.5–5.7), it could be seen that the in-plane properties of the 2.5D extruded cellular structures exhibits higher sensitivity to the relative density, which means that at lower relative densities the structures exhibit significantly weakened mechanical properties. Therefore, when these designs are exploited for energy absorption applications, they are often loaded along in-plane directions Eqs. (5.8 and 5.10), whereas for stiffness reinforcement applications, they are often loaded along out-of-plane directions (Eqs. 5.9 and 5.11).

The selection of unit cell geometry has significant impact on the mechanical performance of the 2.5D extruded cellular structures. Figure 5.55b shows multiple 2D unit cell pattern designs, whose mechanical properties are shown in Fig. 5.56 [157]. Comparing different designs, many structures exhibit rather linear relationship with relative densities except for hexagonal honeycomb, and the diamond design exhibits highest overall performance. On the other hand, for different mechanical property criteria different designs perform differently. For example, the square structure exhibits the highest elastic modulus but very low shear modulus and yield strengths. Some of these differences are more intuitive and easy to explain, such as the low shear modulus of the square structure, while the others are

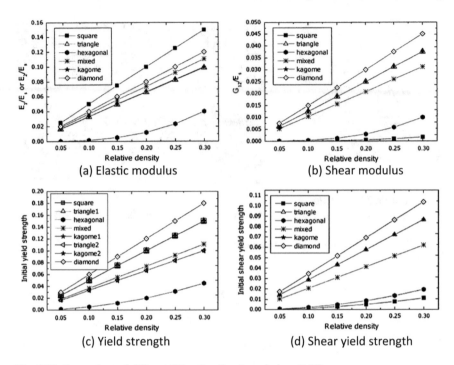

Fig. 5.56 Comparison of different 2D unit cell pattern designs [157]

less obvious, such as the difference between the Kagome structure and the hexagonal honeycomb structure.

One classification method that is widely used in structural analysis could be applied to cellular structure design to help explaining some of the performance difference between different cellular designs. In the determination of structural stability, Maxwell's criterion is often employed, which determines the statical and kinematical determinacy of a pin-jointed frame. For 2D structures, the Maxwell criterion is expressed as [158]:

$$M = b - 2j + 3 \tag{5.12}$$

where M is the Maxwell stability number, and b, j are the number of struts and joints of the framework, respectively. Figure 5.57 demonstrates different 2D framework systems that exhibit different Maxwell stability number. $M < 0$ indicates a kinematically instable framework system (Fig. 5.57a), which upon loading is unable to retain original shape or construction. $M = 0$ indicates a kinematically determinant system with no redundancy (Fig. 5.57b), while $M > 0$ indicates a kinematically over-determinant system (Fig. 5.57c) that requires additional compatibility equations to fully demonstrate. In the case of metal cellular structures, the joints usually exhibit high rigidity and therefore could not be considered as pin

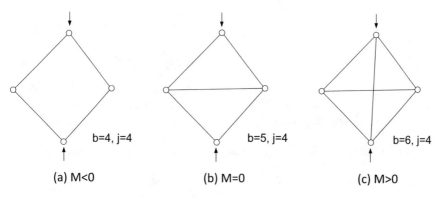

Fig. 5.57 Maxwell stability of different 2D pin-jointed frameworks [158]

joints. However, the Maxwell criterion will determine the dominant deformation mechanism of the structure. For cellular structures with $M < 0$, the dominant deformation mechanism is strut/wall bending, and the structure is termed bending-dominated structure. For cellular structures with $M < 0$, the dominant deformation mechanism is strut/wall stretching/compressing, and therefore the structure is termed stretch-dominated structure. For cellular structures with $M = 0$, a mixed deformation mode is often observed, however the structure could still be considered primarily stretch-dominated.

There exist significant difference of mechanical characteristics between stretch-dominated and bending-dominated cellular structures. Bending-dominated structures usually exhibit lower modulus and strength compared to stretch-dominated structures due to the lower resistance of strut and thin wall structures to bending deformation. As shown in Fig. 5.58a, the bending-dominated structures usually exhibit the classical plateau stress–strain curve, which has been discussed previously in details. In fact, most commercial stochastic cellular structures have Maxwell stability $M < 0$ and exhibit bending-dominated deformation mechanism. In comparison, the stretch-dominated structures exhibit a slightly different stress–strain characteristic. As shown in Fig. 5.58b, the initial failure of stretch-dominated structures usually occurs at higher stress levels. However, upon initial failure the local catastrophic of structure collapse will often cause sudden drop of stress as a result of structural relaxation, which can be termed as post-yield softening. Such structural collapse is also often accompanied by macroscopic geometrical distortion of the structures, which sometimes alter the deformation mechanism of the remaining structures into bending-dominated structures by introducing loading misalignment. The example stress–strain curve shown in Fig. 5.58b demonstrates such case, in which the consequent deformation exhibits plateau stage before densification occurs. However, in some cases the local collapse does not cause the change of deformation mechanism, and consequently the post-yield softening could happen again when another local collapse takes place.

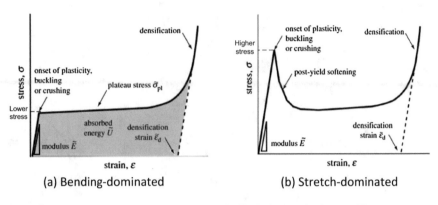

Fig. 5.58   Stress–strain characteristics of different deformation mechanisms [158]

For hexagonal honeycomb structures, the deformation mode under in-plane loading exhibits bending-dominated mechanism, whereas the deformation mode under the out-of-plane loading exhibits stretch-dominated mechanism. Through generalized modeling analysis it could be shown that Eqs. (5.8) and (5.10) could be used to describe the general in-plane properties of bending-dominated structures, while Eqs. (5.9) and (5.11) could be used to describe the general in-plane properties of stretch-dominated structures as well as the out-of-plane properties of both types of cellular structures. Reflecting back about the discussions of different 2D unit cell pattern designs shown in Fig. 5.55b, The square and hexagonal structures are bending-dominated structures, the diamond structure is a stretch-dominated structure with $M = 0$, while the other structures are stretch-dominated structures with $M > 0$. One of the main reasons that the bending-dominated square structure exhibits the highest elastic modulus is due it its efficient alignment of structures along the loading direction, which in effect results in stretch-dominated deformation mode. This also demonstrates another often neglected aspect during the cellular design, which is the alignment between the unit cell orientation and the loading direction. Optimized alignment could result in the strengthening of structural performance, while improperly oriented cellular design might result in significant performance deviation from the designs.

The 2.5D cellular structures could also be designed using analytical modeling approach. Some of the detailed description of the methodology can be found from [146]. For a hexagonal honeycomb unit cell design, there exist four geometric parameters, H, L, θ and t, and two principal directions with structural symmetry which are X_1 and X_2 as shown in Fig. 5.59a. Therefore, the mechanical properties of each axis need to be modeled separately. Note that the extruded length of the cellular structure (b) is not shown, since it always has linear effect on all the in-plane properties and could be simply included into the final equations as a scaling factor.

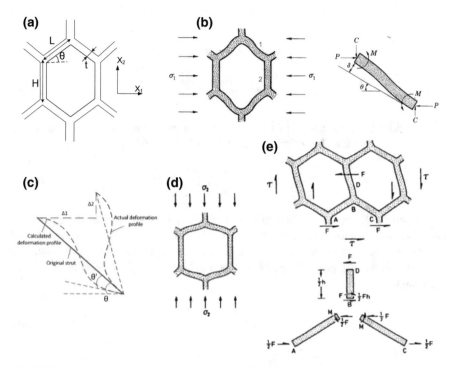

Fig. 5.59 Analytical modeling of hexagonal honeycomb 2.5D extruded cellular structure. **a** Hexagonal honeycomb unit cell. **b** Compression in X_1 direction [146]. **c** Calculation of deformation. **d** Compression in X_2 direction [146]. **e** Shear [146]

First of all, the relative density of the honeycomb structure could be determined by its geometric parameters as:

$$\rho_r = \frac{t}{L} \frac{\alpha + 2}{(\varepsilon - \cos \theta) \sin \theta} \tag{5.13}$$

where $\alpha = H/L$ is the unit cell aspect ratio. Although Eq. (5.13) tends to overestimate the relative densities of honeycomb structures with short and thick walls, it is reasonably accurate for the designs with lower relative densities.

When the honeycomb structure is subjected to uniaxial compressive stress in the X_1 direction as shown in Fig. 5.59b, the cell walls perpendicular to the stress direction (labeled "2") do not contribute to the deformation of the unit. Therefore, the deformation analysis could be carried out with the angled cell walls (labeled "1"). This could be effectively achieved through beam analysis. From force equilibrium and boundary conditions the force components of the angled cell walls as shown in Fig. 5.59b are:

$$C = 0 \tag{5.14a}$$

$$P = \sigma_1 b (H + L \sin \theta) \tag{5.14b}$$

$$M = \frac{PL \sin \theta}{2} \tag{5.14c}$$

Consider the deformation of a beam with both ends constrained by rigid joints, and the overall deflection of the beam could be calculated by:

$$\Delta = \theta' L = \frac{ML^2}{6EI} = \frac{PL^3 \sin \theta}{Ebt^3} \tag{5.15}$$

Figure 5.59c shows the deflected shape of the beam, and due to the rigid joint assumption, the tilt angle of the tangent of the end of the deflected beam should remain constant after deformation. As a result, the actual deflection profile could be determined, and the deformation of the beam in both X_1 and X_2 directions could be determined:

$$\Delta_1 = \theta' L \sin \theta = \frac{PL^3 \sin^2 \theta}{Ebt^3} \tag{5.16a}$$

$$\Delta_2 = \theta' L \cos \theta = \frac{PL^3 \sin \theta \cos \theta}{Ebt^3} \tag{5.16b}$$

When boundary conditions and size effects are both neglected, Eqs. (5.16a, 5.16b) could be used to efficiently predict the total deformation of the honeycomb structure under given compressive stress, therefore estimating both the elastic modulus and Poisson's ratio, which are given as:

$$E_1 = \frac{\sigma_1}{\varepsilon_1} = \frac{\sigma_1 L \cos \theta}{\Delta_1} = \frac{Et^3 \cos \theta}{(H + L \sin \theta) L^2 \sin^2 \theta} \tag{5.17}$$

$$v_{12} = \frac{\cos^2 \theta}{(\alpha + \sin \theta) \sin \theta} \tag{5.18}$$

Using similar methods, the compressive properties in the X_2 direction and the shear properties as shown in Fig. 5.59d, e could be modeled consequently. The elastic modulus and Poisson's ratio in X_2 direction and the shear modulus are given as [146]:

$$E_2 = = \frac{Et^3 (H + L \sin \theta)}{L^3 \cos^3 \theta} \tag{5.19}$$

$$v_{21} = \frac{(\alpha + \sin\theta)\sin\theta}{\cos^2\theta} \tag{5.20}$$

$$G_{12} = \frac{Et^3(\alpha + \sin\theta)}{L^3\alpha^2\cos\theta(1 + 2\alpha)} \tag{5.21}$$

With elastic modulus, shear modulus and Poisson's ratio all known, it is possible to completely determine the deformation of the honeycomb structure under multi-axial loading conditions as long as these loading are applied along principal directions (X_1 and X_2). However, the homogenization treatment that substitute the honeycomb structure with a solid material with equivalent properties predicted by Eqs. (5.17–5.21) might not yield satisfactory design predictions in many cases. With solid material principal stress could be used to uniquely determine the stress status of an infinitesimal element. However, with honeycomb structures, the global and local stress status of the structure exhibit different characteristics when loaded under an "equivalent" case with principal stresses compared to the original loading case. This could be help understood through the fact that honeycomb cellular structures only possess a certain degree of rotational symmetry, which implies that not all orientations can be treated equally as is the case for a continuous solid media. Such argument is visually illustrated in Fig. 5.60, in which the deformation of the honeycomb structure under a combined loading of compression and shear is significantly different from the case with the same structure subject to equivalent principal stress at a rotated orientation according to the principal stress formulation. Note that the deformation in Fig. 5.60 is exaggerated for visual comparison.

In the modeling of the size effect, structural symmetry in relation to the loading stress must be considered. As shown in Fig. 5.61, when the stress is applied along the X_2 direction of the honeycomb structure, the unit cell structure exhibits multiple

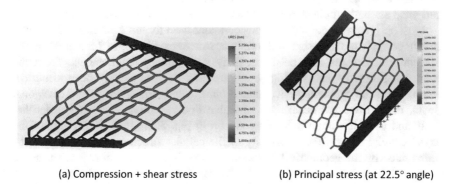

(a) Compression + shear stress   (b) Principal stress (at 22.5° angle)

Fig. 5.60 Comparison of honeycomb structure under two loading cases that are equivalent for continuum

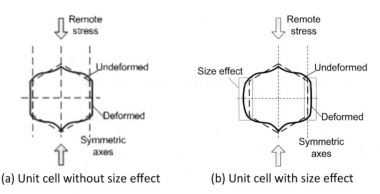

(a) Unit cell without size effect   (b) Unit cell with size effect

Fig. 5.61 Modeling of size effects for honeycomb structure

types of symmetry including mirror symmetry and two-fold rotational symmetry. When the size effect is minimized, the unit cell could be considered to also meet the translational symmetry along X_1 direction, which indicates that the vertical cell walls (i.e. cell walls along X_2 direction) at two sides of the unit cell could only be subjected to axial deformation, which in turn implies that only compressive stress exists in these cell walls. On the other hand, when the unit cell is located at the boundary of a finite structure, since the bending moment is not balanced for the vertical boundary cell wall, the cell wall will be subjected to additional bending as shown in Fig. 5.61b. Analytical expression for each individual unit cell that exhibits size effect could be modeled based on both force equilibrium and the structural compatibility. Analytical solution for size effect with small number of unit cells is feasible, however as the number of unit cell increases, derivation of pure analytical solution becomes computationally inhibitive, and numerical solution might be needed [159]. Theoretical predictions of size effect for honeycomb structure is shown in Fig. 5.62, which shows that the size effect of both modulus and yield strength diminishes at unit cell number approaches 10–12 [159].

Most design methodology and considerations used for the 2.5D extruded cellular structures apply to the design of 3D cellular structures. Similar to 2.5D unit cell, the 3D unit cell designs also need to satisfy space filling requirements. From the elemental geometry theory multiple space filling polyhedral have been identified, among them the commonly encountered ones include triangular prism, hexagonal prism, cube, truncated octahedron, rhombic dodecahedron, elongated dodecahedron, and gyrobifastigum, which are shown in Fig. 5.63 [160]. Regardless of the actual cellular design, the periodicity of the cellular structure must be able to be represented by the geometrical bounding volume in the shape of one of the space filling polyhedral. For example, for the cellular structures shown in Figs. 5.64a and 5.60b, the unit cell bounding volume are cube and hexagonal prism, although the construction of the unit cell within the bounding volume could take more complex designs. As described in Fig. 5.49, different spatial pattern of even the same

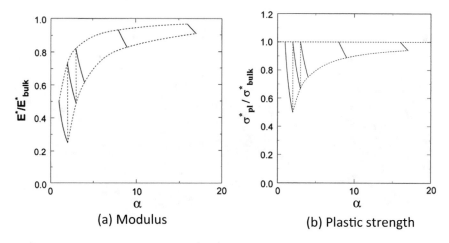

(a) Modulus   (b) Plastic strength

Fig. 5.62 Predicted size effects for hexagonal honeycomb structure [159]

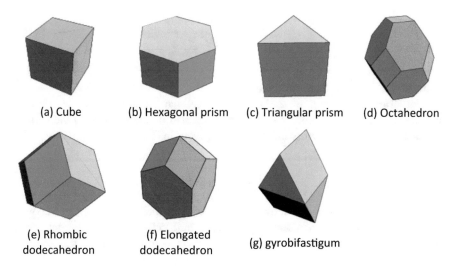

(a) Cube   (b) Hexagonal prism   (c) Triangular prism   (d) Octahedron

(e) Rhombic   (f) Elongated
dodecahedron   dodecahedron   (g) gyrobifastigum

Fig. 5.63 Typical space filling polyhedral [160]

topology design could lead to different unit cell definitions with different geometrical bounding volume. For simple periodic 3D cellular structures with consistent
unit cell shape and dimensions, the relative densities of the structures could be
represented by the relative densities of the unit cell, and therefore the correct
definition of the geometrical bounding volume becomes critical as it determines the
total volume in the calculation. Consequently, the relative density ρ_r could be
determined as:

(b)

(a)

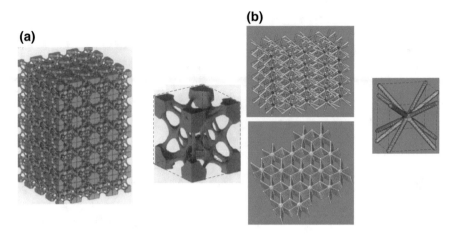

Fig. 5.64 Geometrical bounding volumes for 3D cellular structures. **a** Cubic bounding volume [161]. **b** Hexagonal prism bounding volume

$$\rho_r = \frac{V_s}{V_b} \tag{5.22}$$

where V_s and V_b represent the volume of the solid and geometrical bounding volume of the unit cell, respectively.

Similar to the design of 2.5D extruded cellular structures, the deformation mechanisms of the 3D cellular structures could also be identified with the help of the Maxwell stability criterion, which is given in Eq. (5.23) for 3D cellular structures:

$$M = b - 3j + 6 \tag{5.23}$$

For the unit cell designs shown in Fig. 5.65, as indicated from the Maxwell stability M value, the diamond and rhombic unit cells generally exhibit lower mechanical strength, while the octet truss and octahedron generally exhibit higher

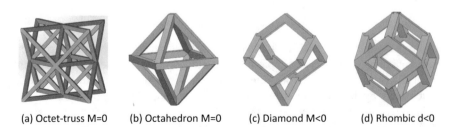

(a) Octet-truss M=0   (b) Octahedron M=0   (c) Diamond M<0   (d) Rhombic d<0

Fig. 5.65 Maxwell stability of different 3D cellular designs

mechanical strength. As a simple rule of thumb, the cellular designs with triangular facets (both external and internal) tend to become stiffer and less compliant. In general the analysis of the deformation mechanism for 3D cellular structures is more complex, and the structures are also more sensitive to the boundary conditions.

It is rather difficult for the 3D cellular structures to achieve near isotropic mechanical properties. Due to the ordered arrangement of discrete strut solids, 3D cellular structures normally possess a finite number of distinct symmetry axes, and their mechanical properties vary when loaded along directions that deviate from these axes even when the number of unit cells are large. On the other hand, identical properties along the principal axes could be achieved, such as the octet truss example given in Fig. 5.65, which has three identical axes aligned orthogonally. In the same sense, it is possible to design for 3D cellular structures that possess different mechanical properties along different loading directions. Such design flexibility renders 3D cellular designs advantageous for applications where multi-axial properties are required.

The analytical design of 3D cellular structure could follow the same approach used for the 2.5D extruded cellular structures. Once the unit cell structure is determined, different boundary conditions could be stablished based on the loading conditions, and the unit cell structures could be further decomposed into individual struts that are fully defined in order to analyze their deformation status. With high degree of structural symmetry the unit cell structures could generally be further simplified into partial structures for the stress analysis. For example, when all the tilting angles and lengths of the struts in a diamond unit cell (Fig. 5.65c) are identical, the struts exhibit identical loading conditions and boundary conditions, and therefore the analysis could be simplified into the modeling of an arbitrary strut in the unit cell structure.

One such example of modeling is shown below in details for the BCC lattice which is shown in Fig. 5.66 [162]. The unit cell geometry for the cellular structure shown in Fig. 5.66a could be represented by either Fig. 5.66b or Fig. 5.66c. However the unit cell layout 2 is easier to model and will therefore be adopted for later modeling analysis.

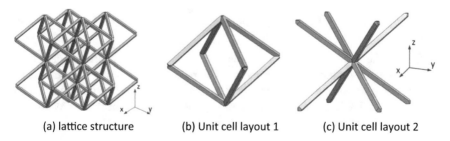

(a) lattice structure   (b) Unit cell layout 1   (c) Unit cell layout 2

Fig. 5.66 Octahedral cellular structure [162]

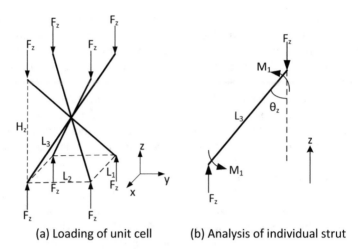

(a) Loading of unit cell   (b) Analysis of individual strut

Fig. 5.67 Octahedral unit cell under compression

From the topology of the unit cell it is obvious that the uniaxial mechanical properties of the BCC lattice could be modeled with identical formulations. Considering a remote compressive stress σ_z applied on the structure along the z-direction, the loading of the unit cell is shown in Fig. 5.67a. Since all struts are subjected to identical loading conditions and boundary conditions, an arbitrary strut is taken for modeling, whose loading condition is illustrated in Fig. 5.67b. From force equilibrium the force components shown in Fig. 5.67b can be determined as:

$$F_z = \frac{1}{4}\sigma_z L_1 L_2 \tag{5.24a}$$

$$M_1 = \frac{1}{2}F_z L_3 \sin\theta_z = \frac{1}{8}\sigma_z L_1 L_2 L_3 \sin\theta_z \tag{5.24b}$$

where L_1, L_2, are the dimensions of the unit cell in x and y directions, and L_3 is the length of a half-strut, and θ_z is the slope angle of the strut in relation to the z axis:

$$\theta_z = \sin^{-1}\frac{\sqrt{L_1^2 + L_2^2}}{2L_3} \tag{5.24c}$$

Under the loading condition shown in Fig. 5.67b, the strut undergoes a bending/shearing combined deformation. For metal cellular struts the axial deformation along the strut axis is relatively insignificant and therefore could be ignored [163]. Employing beam analysis, the deformation of the strut in the z-direction and the direction perpendicular to the z-direction can be obtained as:

$$\Delta z = L_3 \sin \theta_z \left(\frac{M_1 L_3}{6EI} + \frac{6F_z \sin \theta_z}{5GA} \right) \tag{5.25}$$

$$\Delta z_\perp = L_3 \cos \theta_z \left(\frac{M_1 L_3}{6EI} + \frac{6F_z \sin \theta_z}{5GA} \right) \tag{5.26}$$

where E, G are the modulus of elasticity and shear modulus of the solid material, I is the second moment of inertia of bending in the plane shown in Fig. 5.67b, and A is the cross-sectional area. If the deflection $\Delta z_\perp$ is further decomposed into deformations in the x and y directions, Δx and Δy, then it follows that:

$$\Delta x = L_3 \cos \beta_z \cos \theta_z \left(\frac{M_1 L_3}{6EI} + \frac{6F_z \sin \theta_z}{5GA} \right) \tag{5.27}$$

$$\Delta y = L_3 \sin \beta_z \cos \theta_z \left(\frac{M_1 L_3}{6EI} + \frac{6F_z \sin \theta_z}{5GA} \right) \tag{5.28}$$

Therefore, the modulus of the unit cell can be determined as:

$$E_z = \frac{L_3 \cos \theta_z \sigma_z}{\Delta z} = \frac{H_z \sigma_z}{2 \Delta z} = \frac{H_z}{2 L_1 L_2 L_3 \sin^2 \theta \left(\frac{L_3^2}{48EI} + \frac{3}{10GA} \right)} \tag{5.29}$$

where $H_z = 2L_3 \cos \theta = 2 \sqrt{L_3^2 - \frac{1}{4}(L_1^2 + L_2^2)}$, which is also shown in Fig. 5.67a.

The strength of the BCC lattice could be roughly estimated by the initial yield strength of the strut. Such conservative estimation could sometimes be justified by the fact that metal AM cellular structures often exhibit significant quality variability. From Fig. 5.67b the normal stress and shear stress of the strut could be determined as:

$$\sigma_1 = \frac{M_1 u}{I} + \frac{F_z \cos \theta_z}{A} = \frac{\sigma_z L_1 L_2 L_3 \sin \theta_z}{8I} u + \frac{\sigma_z L_1 L_2 \cos \theta_z}{4A} \tag{5.30a}$$

$$\tau_1 = \frac{F_z \sin \theta_z D}{Ib} \tag{5.30b}$$

where u is the distance from the geometrical center of the cross section to the location of interest, D is the moment of area of the cross section, and b is the width of the cross section at the location of interest. Apply Von Mises Criterion, the maximum allowable stress level σ_m that results in the onset of structure yield is:

$$\sigma_m = \frac{4}{L_1 L_2 \sqrt{\frac{L_3^2 \sin^2 \theta_z}{2I^2} u^2 + \frac{2 \cos^2 \theta_z}{A^2} + \frac{6 \sin^2 \theta_z D^2}{I^2 b^2}}} \sigma_Y \tag{5.31}$$

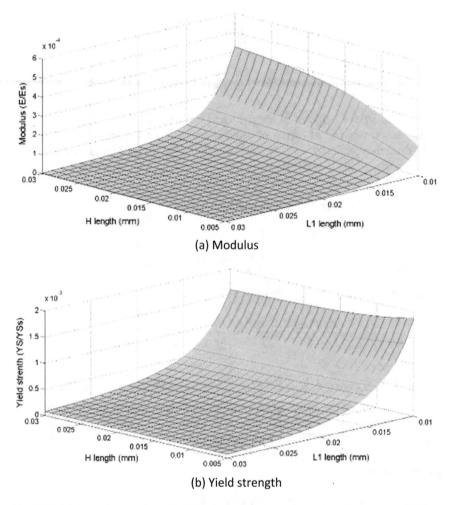

(a) Modulus

(b) Yield strength

Fig. 5.68 Mechanical properties of BCC lattice as functions of geometrical parameters [162]

Following Eqs. (5.29 and 5.31), through geometrical parameter design a range of mechanical properties could be achieved in particular direction of the BCC lattice as shown in Fig. 5.68. Although the mechanical properties of the BCC lattice along the three principal directions are not independent, due to the broad range of properties achievable, it is still possible to achieve multi-axial optimization.

An interesting characteristic of the BCC lattice is its reversed size effect. From previous conversations it is known that for most cellular structures the elastic modulus and strength increases with increasing unit cell numbers until the size effect diminishes. However for the BCC lattice, the increase of unit cell numbers along the compression direction results in decrease of its mechanical properties.

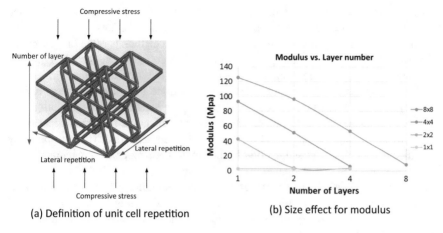

(a) Definition of unit cell repetition   (b) Size effect for modulus

Fig. 5.69 Reversed size effect for BCC lattice [162]

As shown in Fig. 5.69, with only one layer of unit cell, the compressive modulus of the BCC lattice increases with increasing number of unit cells in the lateral direction. As the number of layer increases, the elastic modulus of the BCC lattice reduces by following a logarithmic law, and when the number of layers is sufficiently large, the size effect also appears to diminish at low level of modulus. This rather counterintuitive behavior is a result of the boundary constraint, which causes stress concentration in the structures and therefore reduction of mechanical properties. Such reversed size effect could possess significant practical values in the design of sandwich panels, which indicates that a small number of layers could be designed for the sandwich structures to achieve high mechanical properties.

Similar to 2.5D extruded cellular structures, most 3D cellular structures lack sufficient structural symmetry for the homogenization to be applicable except for limited loading cases. Interested readers could refer to published literature for additional information [164].

5.6.2   Material/Process Design for Lightweight Structures

So far the discussions of AM lightweight structure design have been largely focused on geometric design. It is, however, important to note that the process design also plays a critical role. In general the manufacturing of cellular structures is more challenging from quality control perspective due to the fact that the dimensions of the cellular features often approach the limit of the systems. For systems such as electron beam melting (EBM) such limitation is especially significant, since the process currently has larger energy beam size compared to the laser based systems. It is generally perceived that the current triangular tessellation based. STL file format introduces error into the geometrical representations, however not such observation has been reported particularly for cellular structures. In addition, the adoption of more accurate geometrical representation method is often limited by the lack of machine adoption and potentially intensive computation costs [165].

Limited literatures are available for the investigation of manufacturing issues with cellular structures, although it has been identified that factors such as energy density, energy beam power, scanning speed, part location, part orientation and scanning strategies all have potentially significant effect on the mechanical properties of the cellular structures [140, 166–169]. Staircase effect tends to be more significant for cellular structures due to the small dimensions of the geometries. It has been suggested that at smaller orientation angles (i.e. more aligned to horizontal plane) the staircase effect of struts could become significant enough to affect the structural integrity as shown in Fig. 5.70a [170]. As shown in Fig. 5.70b–c, at 20° the cross section geometry of the strut exhibits more significant fluctuation compared to the 70° struts [171].

Beside staircase effect, another intrinsic effect that introduces geometrical error is the surface powder sintering for the powder bed fusion AM processes, which is caused by the heat dissipated away from the processed areas. As shown in Fig. 5.71, these surface defects causes dimensional variations on the cellular struts. For the calculation of mechanical properties, it was suggested that the minimum strut dimension (d in Fig. 5.71b) should be used in the modeling [172]. On the other hand, in order to calculate pore size, the largest strut dimension (D in Fig. 5.71b) should be adopted. For bulky structures such surface roughness could potentially reduce fatigue performance, and for cellular structures, such effect could be more pronounced due to the large specific surface areas of these structures. Literatures have shown that the fatigue strength of Ti6Al4V AM cellular structures are generally lower compared to that of the bulky Ti6Al4V fabricated via the same AM processes [116, 140, 173, 174]. Due to the complex geometries and the extensive existence of internal features, cellular structures are generally difficult to perform surface treatment with. As a result, the fatigue performance of the metal AM cellular structures still poses a significant barrier for their structural applications.

The microstructure of the metal cellular structures fabricated via powder bed fusion AM often exhibits finer microstructural phases compared to the bulky

(a)   **(b)**   **(c)**

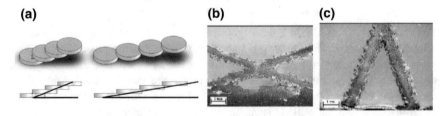

Fig. 5.70 Staircase effect for cellular structures. **a** Staircase effect with different orientation angles [170]. **b** struts at 20° angle [171]. **c** struts at 70° angle [171]

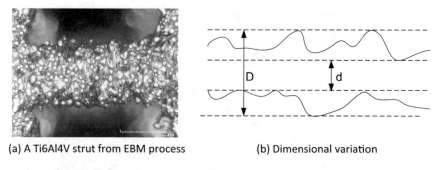

(a) A Ti6Al4V strut from EBM process   (b) Dimensional variation

Fig. 5.71 Surface quality issue with thin struts [172]

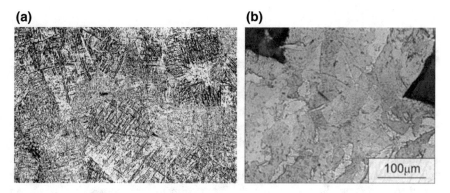

Fig. 5.72 Microstructure of AM Ti6Al4V cellular struts. **a** Electron beam melting [175]. **b** Laser melting [169]

structures. As shown in Fig. 5.72, for Ti6Al4V cellular structures fabricated via electron beam and laser beam based powder bed fusion AM processes, the predominant microstructure is the fine-grained lath α' martensite, which is suggested to be contributed by the rapid cooling effect introduced by large surface area of the cellular geometries [169, 175]. In addition, it was observed that the size of the prior β grains in the microstructure is also a function of both strut orientation and strut dimensions [176]. As shown in Fig. 5.73, for Ti6Al4V thin struts fabricated by laser melting process, with increasing feature dimensions, the microstructural grain size of the Ti6Al4V struts exhibit an increasing trend, while there also exist significant size differences between grains that are close to the exterior surface of the struts and those that are at the interior of the struts. On the other hand, smaller overhanging angle result in more consistent grain size distribution throughout the strut. Such microstructural variation likely corresponds to local mechanical property variations, which means that the mechanical properties of the cellular structures are likely coupled with their geometrical designs and process setup. However, so far no

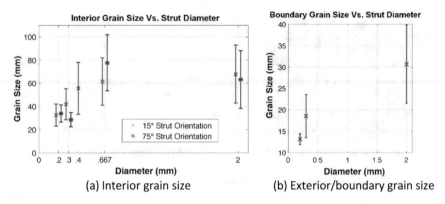

Fig. 5.73 Grain size dependency on orientation and strut diameter for Ti6Al4V fabricated by laser melting [176]

such knowledge exists, which means that currently the only available method to account for the material effect is to perform experimental testing with the fabricated cellular structures.

Another interesting process design aspect is the selection of scanning strategies. Although such parameter selection is often dependent on individual process systems, the designers should be aware of its potential impact of the manufacturability. For example, in the EOS metal laser melting system, there exist several options for scanning strategies, including the contour, the hatch and the edge strategies. For each material, there also exist various beam offset setting to compensate for the characteristic melting pool diameter. When the feature dimension becomes small as shown in Fig. 5.74, the actual scanning paths become more sensitive to the selection of scanning strategies and the setting of offset [173]. For a particular material with beam offset value of 0.05 mm, the characteristic beam diameter could be considered

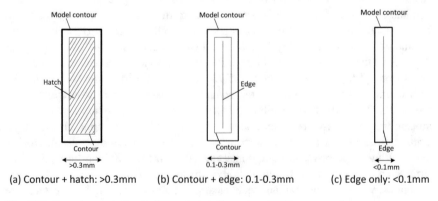

Fig. 5.74 Scanning strategies at different strut dimensions [173]

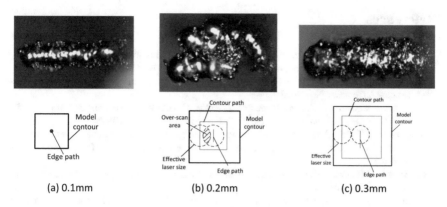

Fig. 5.75 Scanning path overlap for different features using contour+edge scanning [173]

as 0.1 mm (i.e. twice the value of the beam offset). Therefore, the contour + hatch scanning strategy shown in Fig. 5.74a could effectively create a scanning pattern that fills the entire cross-sectional area without having scanning track overlap for struts that are thicker than 0.3 mm. However, when the strut dimension falls between 0.1 and 0.3 mm, no scanning strategy combinations could generate fulfilling scanning path without scanning track overlap. When the dimension of the strut feature is smaller than 0.1 mm, regardless of the selection of scanning strategies, only edge scanning strategy would be possible, which generate a single-pass scanning track across the cross sections regardless of the designed thickness. As a result, the actual feature size for all the struts with dimensions smaller than 0.1 mm will be identical throughout the cellular structure and becomes only dependent on the other non-geometrical parameters such as power and scanning speed.

As a general guideline, during the fabrication of cellular structures scanning track overlap should be avoided. Experimental study showed that repeated heating in the scanned areas often result in fabrication failure for cellular strut features. As shown in Fig. 5.75, when contour + edge scanning strategy is adopted for the fabrication of cellular structures with the materials that has 0.1 mm beam offset in EOS M270 metal system, the struts with 0.2 mm cross-sectional dimension would result in maximum scanning track overlap. Consequently, the fabrication of these struts either exhibits significant errors as shown in Fig. 5.75b or completely fails to build [173].

In conclusion, it can be seen that the knowledge for AM lightweight structure design is rather fragmented now. Relatively in-depth studies have been carried out with either the geometrical design or the material properties, however little knowledge is currently available on how these two design aspects could be effectively integrated. A regularly employed shortcut is to perform experimental-based structure characterization which, if planned with the knowledge introduced in this chapter in mind, could result in efficient development of structural design guidelines for that particular cellular designs.

References

1. http://www.geaviation.com/company/additive-manufacturing.html. Accessed Nov 2015
2. http://www.aerojet.org/additive-manufacturing. Accessed Nov 2015
3. http://today.uconn.edu/2013/04/pratt-whitney-additive-manufacturing-innovation-center-opens-at-uconn/. Accessed Nov 2015
4. http://www.bonezonepub.com/component/content/article/689-48-bonezone-march-2013-research-a-development-the-future-of-additive-manufacturing-in-orthopaedic-implants
5. Fukuda H (2015). Additive manufacturing technology for orthopedic implants. In: Advances in metallic biomaterials. Springer Series in Biomaterials Science and Engineering, vol 4. pp 3–25
6. Industry 4.0 (2015) The future of productivity and growth in manufacturing industries. The Boston Consulting Group
7. Industry 4.0 (2014) Challenges and solutions for the digital transformation and use of exponential technologies. Deloitte
8. http://www.ifam.fraunhofer.de/en/Bremen/Shaping_Functional_Materials/Powder_Technology/Additive_Manufacturing.html. Accessed Nov 2015
9. McKee C (2015) Design for additive manufacturing—DMLS the designers mindset. In: Additive Manufacturing Users Group (AMUG) Annual Conference. Jacksonville
10. Stucker B (2015) Support optimization for metal laser sintering. RAPID, Long Beach
11. Arcam material specification. http://www.arcam.com/technology/products/metal-powders/. Accessed Nov 2015
12. EOS metal materials specification. http://www.eos.info/material-m. Accessed Nov 2015
13. SLM Solutions. http://stage.slm-solutions.com/index.php?low-melting-alloy-applications_en. Accessed Nov 2015
14. D Systems ProX specification. http://www.3dsystems.com/3d-printers/production/prox-300. Accessed Nov 2015
15. Renishaw material specification. http://www.renishaw.com/en/data-sheets-additive-manufacturing–17862. Accessed Nov 2015
16. Xing X, Yang L (2015) A glance at the recent additive manufacturing research and development in China. In: Proceedings of the Solid Freeform Fabrication (SFF) Symposium, Austin, TX, USA
17. Sciaky additive manufacturing. http://www.sciaky.com/additive-manufacturing/wire-am-vs-powder-am
18. MarkForged materials. https://markforged.com/materials/. Accessed Nov 2015
19. Stratasys materials. http://www.stratasys.com/materials/fdm. Accessed Nov 2015
20. D Systems polymer materials specification. http://www.3dsystems.com/materials/production. Accessed Nov 2015
21. EOS polymer materials. http://www.eos.info/material-p. Accessed Nov 2015
22. PLA typical properties. http://plastics.ulprospector.com/generics/34/c/t/polylactic-acid-pla-properties-processing
23. McLaws IJ (1971) Uses and specification of silica sand. Research Council of Alberta
24. D Systems ProJet 160 Specification. http://www.3dsystems.com/3d-printers/personal/projet-160. Accessed Nov 2015
25. Lithoz Materials. http://www.lithoz.com/en/products/materials/. Accessed Nov 2015
26. Robocasting. https://en.wikipedia.org/wiki/Robocasting. Accessed Nov 2015
27. Calcium hydroxylapatite. http://matweb.com/search/DataSheet.aspx?MatGUID=e1654c43ab994d7fab5e0f9aabe4dddc&ckck=1
28. Lithoz Product Gallery. http://www.lithoz.com/en/picture-gallery/startseite/. Accessed Nov 2015
29. Robocasting Filtration Products. https://robocasting.net/index.php/products/filtration. Accessed Nov 2015

30. Stratasys Connex 3. http://www.stratasys.com/3d-printers/production-series/connex3-systems. Accessed Nov 2015
31. D Systems ProJet 3500. http://www.3dsystems.com/3d-printers/professional/projet-3500-hdmax. Accessed Nov 2015
32. Reeves PE, Cobb RC (1998). Reducing the surface deviation of stereolithography using an alternative build strategy. In: Proceedings of Solid Freeform Fabrication (SFF) Symposium, Austin, TX, USA
33. German RM (1984) Powder Metallurgy Science, Metal Powder Industries Federation
34. Boivie K (2001) Limits of loose metal powder density in the Sinterstation. In: Proceedings of Solid Freeform Fabrication (SFF) Symposium, Austin, TX, USA
35. Yang L, Anam MA (2014) An investigation of standard test part design for additive manufacturing. In: Proceedings of Solid Freeform Fabrication (SFF) Symposium, Austin, TX, USA
36. Ippolito R, Iuliano L, Gatto A (1995) Benchmarking of rapid prototyping techniques in terms of dimensional accuracy and surface finish. Ann CIRP 44:157–160
37. Childs THC, Juster NP (1994) Linear and geometric accuracies from layer manufacturing. Ann CIRP 43:163–166
38. Mahesh M, Wong YS, Fuh JYH, Loh HT (2004) Benchmarking for comparative evaluation of RP systems and processes. Rapid Prototyping J 10:123–135
39. Kim GD, Oh YT (2008) A benchmark study on rapid prototyping processes and machines: quantitative comparisons of mechanical properties, accuracy, roughness, speed, and material cost. Proc Inst Mech Eng Part B: J Eng Manufact 222:201–215
40. Castillo L (2005) Study about the rapid manufacturing of complex parts of stainless steel and titanium. TNO Industrial Technology
41. Fahad M, Hopkinson N (2012) A new benchmarking part for evaluating the accuracy and repeatability of additive manufacturing (AM) processes. In: Proceedings of 2nd international conference on Mechanical, Production and Automobile Engineering, Singapore
42. Abdelrahman M, Starr TL (2014) Layerwise Monitoring of Polymer Laser Sintering Using Thermal Imaging. In: Proceedings of Solid Freeform Fabrication (SFF) Symposium, Austin, TX, USA
43. Rafi HK, Karthik NV, Starr TL, Stucker BE (2012) Mechanical property evaluation of Ti-6Al-4V parts made using electron beam melting. In: Proceedings of Solid Freeform Fabrication (SFF) Symposium, Austin, TX, USA
44. Moylan S, Slotwinski J, Cooke A, Jurrens K, Donmez MA (2012) Proposal for a standard test artifact for additive manufacturing machines and processes. In: Proceedings of Solid Freeform Fabrication (SFF) Symposium, Austin, TX, USA
45. DM3D technology. http://www.dm3dtech.com/index.php?option=com_content&view=article&id=86&Itemid=552. Accessed Nov 2015
46. Nassar AR, Starr B, Reutzel EW (2015) Process monitoring of directed-energy deposition of Inconel-718 via plume imaging. In: Proceedings of Solid Freeform Fabrication (SFF) Symposium, Austin, TX, USA
47. Abdelrahman M, Starr TL (2015) Feedforward control for polymer laser sintering process using part geometry. In: Proceedings of Solid Freeform Fabrication (SFF) Symposium, Austin, TX, USA
48. Karnati S, Matta N, Sparks T, Liou F (2013) Vision-based process monitoring for laser metal deposition processes. In: Proceedings of Solid Freeform Fabrication (SFF) Symposium, Austin, TX, USA
49. Wroe W, Gladstone J, Phillips T, McElroy A, Fish S, Beaman J (2015) In-situ thermal image correlation with mechanical properties of nylon-12 in SLS. In: Proceedings of Solid Freeform Fabrication (SFF) Symposium, Austin, TX, USA
50. Cloots M, Spierings AB, Wegener K (2013) Assessing new support minimizing strategies for the additive manufacturing technology. In: Proceedings of Solid Freeform Fabrication (SFF) Symposium, Austin, TX, USA

51. Calignano F (2014) Design optimization of supports for overhanging structures in aluminum and titanium alloys by selective laser melting. Mater Des 64:203–213

52. Krol TA, Zaeh MF, Seidel C (2012) Optimization of supports in metal-based additive manufacturing by means of finite element models. In: Proceedings of Solid Freeform Fabrication (SFF) Symposium, Austin, TX, USA

53. Poyraz O, Yasa E, Akbulut G, Orhangul A, Pilatin S (2015) Investigation of support structures for direct metal laser sintering (DMLS) of IN625 parts. In: Proceedings of Solid Freeform Fabrication (SFF) Symposium, Austin, TX, USA

54. Strano G, Hao L, Everson RM, Evans KE (2013) A new approach to the design and optimisation of support structures in additive manufacturing. Int J Adv Manuf Technol 66:1247–1254

55. DSIM News. http://3dsim.com/news/. Accessed Dec 2015

56. Autodesk Meshmixer. http://www.meshmixer.com/index.html. Accessed Dec 2015

57. Yan W, Smith J, Ge W, Lin F, Liu W (2015) Multiscale modeling of electron bean and substrate interaction: a new heat source model. Comput Mech 56:265–276

58. Kruth JP, Wang X, Laoui T, Froyen L (2003) Lasers and materials in selective laser sintering. Assembly Autom 23(4):357–371

59. Abd-Elghany K, Bourell DL (2012) Property evaluation of 304L stainless steel fabricated by selective laser melting. Rapid Prototyping J 18(5):420–428

60. Spierings AB, Herres N, Levy G (2010) Influence of the particle size distribution on surface quality and mechanical properties in additive manufactured stainless steel parts. In: Proceedings of Solid Freeform Fabrication (SFF) Symposium, Austin, TX, USA

61. Spierings AB, Levy G (2009) Comparison of density of stainless steel 316L parts produced with selective laser melting using different powder grades. In: Proceedings of Solid Freeform Fabrication (SFF) Symposium, Austin, TX, USA

62. Karapatis NP, Egger G, Gygax P-E, Glardon R (1999) Optimization of powder laser density in selective laser sintering. In: Proceedings of Solid Freeform Fabrication (SFF) Symposium, Austin, TX, USA

63. Konrad C, Zhang Y, Shi Y (2007) Melting and solidification of a subcooled metal powder particle subjected to nanosecond laser heating. Int J Heat Mass Transf 50:2236–2245

64. Liu B, Wildman R, Tuck C, Ashcroft I, Hague R (2011) Investigation the effect of particle size distribution on processing parameters optimisation in selective laser melting. In: Proceedings of Solid Freeform Fabrication (SFF) Symposium, Austin, TX, USA

65. Shi Y, Zhang Y (2006) Simulation of random packing of spherical particles with different size distributions. In: ASME International Mechanical Engineering Congress, Heat Transfer, Chicago, IL

66. Khairallah SA, Anderson A (2014) Mesoscopic simulation model of selective laser melting of stainless steel powder. J Mater Process Technol 214:2627–2636

67. Korner C, Bauereiβ A, Attar E (2013) Fundamental consolidation mechanisms during selective beam melting of powders. Model Simul Mater Sci Eng 21(085001):1–18

68. Korner C, Attar E, Heinl P (2011) Mesoscopic simulation of selective beam melting processes. J Mater Process Technol 211:978–987

69. King W, Anderson AT, Ferencz RM, Hodge NE, Kamath C, Khairallah SA (2015) Overview of modelling and simulation of metal powder bed fusion process at Lawrence Livermore National Laboratory. Mater Sci Technol 31(8):957–968

70. Kruth J-P, Levy G, Klocke F, Childs THC (2007) Consolidation phenomena in laser and powder-bed based layered manufacturing. CIRP Ann Manufact Technol 56(2):730–759

71. Tolochko NK, Mozzharov SE, Yadroitsev IA, Laoui T, Froyen L, Titov VI, Ignatiev MB (2004) Balling process during selective laser treatment of powders. Rapid Prototyping J 10(2):78–87

72. Kruth JP, Froyen L, Van Vaerenbergh J, Mercelis P, Rombouts M, Lauwers B (2004) Selective laser melting of iron-based powder. J Mater Process Technol 149:616–622

73. Akhtar S, Wright CS, Youseffi M, Hauser C, Childs THC, Taylor CM, Baddrossamay M, Xie J, Fox P, O'Neill W (2003) Direct selective laser sintering of tool steel powders to high density: part B—the effect on microstructural evolution. In: Proceedings of Solid Freeform Fabrication (SFF) Symposium, Austin, TX, USA

74. Attar E (2011) Simulation of selective electron beam melting processes. Ph.D. dissertation, University of Erlangen-Nurnberg

75. Mills KC, Keene BJ, Brooks RF, Shirali A (1998) Marangoni effects in welding. Philos Trans: Math Phys Eng Sci 356:911–925

76. Xiao B, Zhang Y (2007) Marangoni and buoyancy effects on direct metal laser sintering with a moving laser beam. Numer Heat Transfer, Part A 51:715–733

77. Antony K, Arivazhagan N (2015) Studies on energy penetration and Marangoni effect during laser melting process. J Eng Sci Technol 10(4):509–525

78. Louvis E, Fox P, Sutcliffe CJ (2011) Selective laser melting of aluminium components. J Mater Process Technol 211:275–284

79. Aboulkhair NT, Everitt NM, Ashcroft I, Tuck C (2014) Reducing porosity in AlSi10Mg parts processed by selective laser melting. Addit Manufact 1–4:77–86

80. Mumtaz KA, Hopkinson N (2010) Selective laser melting of thin wall parts using pulse shaping. J Mater Process Technol 210:279–287

81. King WE, Barth HD, Castillo VM, Gallegos GF, Gibbs JW, Hahn DE, Kamath C, Rubenchik AM (2014) Observation of keyhole-mode laser melting in laser powder-bed fusion additive manufacturing. J Mater Process Technol 214:2915–2925

82. Gockel J, Beuth J (2013) Understanding Ti-6Al-4V microstructure control in additive manufacturing via process maps. In: Proceedings of Solid Freeform Fabrication (SFF) Symposium, Austin, TX, USA

83. Beuth J, Fox J, Gockel J, Montgomery C, Yang R, Qiao H, Soylemez E, Reeseewatt P, Anvari A, Narra S, Klingbeil N (2013) Process mapping for qualification across multiple direct metal additive manufacturing processes. In: Proceedings of Solid Freeform Fabrication (SFF) Symposium, Austin, TX, USA

84. Cheng B, Chou K (2015) Melt pool evolution study in selective laser melting. In: Proceedings of Solid Freeform Fabrication (SFF) Symposium, Austin, TX, USA

85. Hauser C, Childs THC, Taylor CM, Badrossamay M, Akhtar S, Wright CS, Youseffi M, Kie J, Fox P, O'Neill W (2003) Direct selective laser sintering of tool steel powders to high density: Part A—Effects of laser beam width and scan strategy. In: Proceedings of Solid Freeform Fabrication (SFF) Symposium, Austin, TX, USA

86. Clijsters S, Craeghs T, Kruth JP (2011) A priori process parameter adjustment for SLM process optimization. In: Bartolo PJ et al (eds) Innovative developments in virtual and physical prototyping. CRC Press-Taylor & Francis Group, London, pp 553–560

87. Yadroitsev I, Bertrand Ph, Smurov I (2007) Parametric analysis of the selective laser melting process. Appl Surf Sci 253:8064–8069

88. Yang L, Gong H, Dilip S, Stucker B (2014) An investigation of thin feature generation in direct metal laser sintering systems. In: Proceedings of Solid Freeform Fabrication (SFF) Symposium, Austin, TX, USA

89. Simchi A, Pohl H (2003) Effects of laser sintering processing parameters on the microstructure and densification of iron powder. Mater Eng A 359:119–128

90. Price S, Cooper K, Chou K (2012) Evaluations of temperature measurements by near-infrared thermography in powder-based electron-beam additive manufacturing. In: Proceedings of Solid Freeform Fabrication (SFF) Symposium, Austin, TX, USA

91. Neugebauer F, Keller N, Ploshikhin V, Feuerhahn F, Kohler H (2014) Multi scale FEM simulation for distortion calculation in additive manufacturing of hardening stainless steel. In: International Workshop on Thermal Forming and Welding Distortion, Bremen, Germany

92. Zhang S, Dilip S, Yang L, Miyanaji H, Stucker B (2015) Property evaluation of metal cellular strut structures via powder bed fusion AM. In: Proceedings of Solid Freeform Fabrication (SFF) Symposium, Austin, TX, USA

93. Kasperovich G, Hausmann J (2015) Improvement of fatigue resistance and ductility of TiAl6V4 processed by selective laser melting. J Mater Process Technol 220:202–214

94. Vrancken B, Thijs L, Kruth J-P, Van Humbeeck J (2012) Heat treatment of Ti6Al4V produced by selective laser melting: microstructure and mechanical properties. J Alloy Compd 54:177–185

95. Vandenbroucke B, Kruth J-P (2007) Selective laser melting of biocompatible metals for rapid manufacturing of medical parts. Rapid Prototyping J 13(4):196–203

96. Facchini L, Magalini E, Robotti P, Molinari A (2009) Microstructure and mechanical properties of Ti-6Al-4V produced by electron beam melting of pre-alloyed powders. Rapid Prototyping J 15(3):171–178

97. Attar H, Calin M, Zhang LC, Scudino S, Eckert J (2014) Manufacture by selective laser melting and mechanical behavior of commercially pure titanium. Mater Sci Eng, A 593: 170–177

98. Manfredi D, Calignano F, Krishnan M, Canali R, Ambroso EP, Atzeni E (2013) From powders to dense metal parts: characterization of a commercial AiSiMg alloy processed through direct metal laser sintering. Materials 6:856–869

99. Kempen K, Thijs L, Van Humbeeck J, Kruth J-P (2012) Mechanical properties of AlSi10Mg produced by selective laser melting. Phys Proc 39:439–446

100. Vilaro T, Colin C, Bartout JD, Naze L, Sennour M (2012) Microstructural and mechanical approaches of the selective laser melting process applied to a nickel-base superalloy. Mater Sci Eng, A 534:446–451

101. Amato KN, Gaytan SM, Murr LE, Martinez E, Shindo PW, Hernandez J, Collins S, Medina F (2012) Microstructures and mechanical behavior of Inconel 718 fabricated by selective laser melting. Acta Mater 60:2229–2239

102. Wang Z, Guan K, Gao M, Li X, Chen X, Zeng X (2012) The microstructure and mechanical properties of deposited-IN718 by selective laser melting. J Alloy Compd 513:518–523

103. Paul CP, Ganesh P, Mishra SK, Bhargava P, Negi J, Nath AK (2007) Investigating laser rapid manufacturing for Inconel-625 components. Opt Laser Technol 39:800–805

104. Yadroitsev I, Pavlov M, Bertrand Ph, Smurov I (2009) Mechanical properties of samples fabricated by selective laser melting. In: 14th European Meeting of Prototyping and Rapid Manufacturing, Paris, France

105. Murr LE, Martinez E, Gaytan SM, Ramirez DA, Machado BI, Shindo PW, Martinez JL, Medina F, Wooten J, Ciscel D, Ackelid U, Wicker RB (2011) Microstructrual architecture, microstructures, and mechanical properties for a nickel-base superalloy fabricated by electron beam melting. Metall Mater Trans A 42:3491–3508

106. Gaytan SM, Murr LE, Martinez E, Martinez JL, Machado BI, Ramirez DA, Medina F, Collins S, Wicker RB (2010) Comparison of microstructures and mechanical properties for solid cobalt-base alloy components and biomedical implant prototypes fabricated by electron beam melting. In: Proceedings of Solid Freeform Fabrication (SFF) Symposium, Austin, TX, USA

107. Takachi A, Suyalatu, Nakamoto T, Joko N, Nomura N, Tsutsumi Y, Migita S, Doi H, Kurosu S, Chiba A, Wakabayashi N, Igarashi Y, Hanawa T (2013) Microstructures and mechanical properties of Co-29Cr-6Mo alloy fabricated by selective laser melting process for dental applications. J Mech Behav Biomed Mater 21:67–76

108. Song C, Yang Y, Wang Y, Wang D, Yu J (2014) Research on rapid manufacturing of CoCrMo alloy femoral component based on selective laser melting. Int J Adv Manufact Technol 75:445–453

109. Wu L, Zhu H, Gai X, Wang Y (2014) Evaluation of the mechanical properties and porcelain bond strength of cobalt-chromium dental alloy fabricated by selective laser melting. J Prosthet Dent 111(1):51–55

110. Sun S-H, Koizumi Y, Kurosu S, Li Y-P, Matsumoto H, Chiba A (2014) Build direction dependence of microstructure and high-temperature tensile property of Co-Cr-Mo alloy fabricated by electron beam melting. Acta Mater 64:154–168

111. Biamino S, Penna A, Ackelid U, Sabbadini S, Tassa O, Fino P, Pavese M, Gennaro P, Badini C (2011) Electron beam melting of Ti-48Al-2Cr-2Nb alloy: microstructure and mechanical properties investigation. Intermetallics 19:776–781

112. Facchini B, Vicente N Jr, Lonardelli I, Magalini E, Robotti P, Molinari A (2010) Metastable austenite in 17-4 precipitation-hardening stainless steel produced by selective laser melting. Adv Eng Mater 12(3):184–188

113. Rafi HK, Pal D, Patil N, Starr TL, Stucker BE (2014) Microstructure and mechanical behavior of 17-4 precipitation hardenable steel processed by selective laser melting. J Mater Eng Perform 23(12):4421–4428

114. Hrabe NW, Heinl P, Flinn B, Korner C, Bordia RK (2011) Compression-compression fatigue of selective electron beam melted cellular titanium (Ti-6Al-4V). J Biomed Mater Res Part B Appl Biomater 99:313–320

115. Yavari SA, Wauthle R, van der Stok J, Riemslag AC, Janssen M, Mulier M, Kruth JP, Schrooten J, Weinans H, Zadpoor AA (2013) Fatigue behavior of porous biomaterials manufactured using selective laser melting. Mater Sci Eng, C 33:4849–4858

116. Li SJ, Murr LE, Cheng XY, Zhang ZB, Hao YL, Yang R, Medina F, Wicker RB (2012) Compression fatigue behavior of Ti-6Al-4V mesh arrays fabricated by electron beam melting. Acta Mater 60:793–802

117. Gong H, Rafi K, Starr T, Stucker B (2012) Effect of defects on fatigue tests of as-built Ti-6Al-4V parts fabricated by selective laser melting. In: Proceedings of the Solid Freeform Fabrication (SFF) Symposium, Austin, TX, USA

118. Li P, Warner DH, Fatemi A, Phan N (2016) Critical assessment of the fatigue performance of additively manufactured Ti-6Al-4V and perspective for future research. Int J Fatigue 85:130–143

119. Zein I, Hutmacher DW, Tan KC, Teoh SH (2002) Fused deposition modeling of novel scaffold architectures for tissue engineering applications. Biomaterials 23(4):1169–1185

120. Rosen DW (2007) Design for additive manufacturing: a method to explore unexplored regions of the design space. In: Proceedings of the Solid Freeform Fabrication (SFF) Symposium, Austin, TX

121. Castilho M, Dias M, Gbureck U, Groll J, Fernandes P, Pires I, Gouveia B, Rodrigues J, Vorndran E (2013) Fabrication of computationally designed scaffolds by low temperature 3D printing. Biofabrication 5(3):035012

122. Kuhn R, Minuzzi RFB (2015) The 3d printing's panorama in fashion design. In: Proceedings of 5th Documenta Fashion Seminar and 2nd International Congress of Memory, Design and Fashion, Sao Paulo, Brazil

123. Wang BZ, Chen Y (2014) The effect of 3D printing technology on the future fashion design and manufacturing. Appl Mech Mater 496–500:2687–2691

124. http://www.3ders.org/articles/20120821-continuum-fashion-launches-custom-3d-printed-shoes.html

125. Francis Bitonti Studio. http://www.francisbitonti.com/fiber-tables/

126. Neri Oxman: Projects. http://www.materialecology.com/projects

127. Wake Forest Institute of Regenerative Medicine. http://www.wakehealth.edu/WFIRM/

128. Chahine G, Smith P, Kovacevic R (2010) Application of topology optimization in modern additive manufacturing. In: Proceedings of Solid Freeform Fabrication (SFF) Symposium, Austin, TX

129. Bendsoe MP, Sigmund O (2003) Topology optimizaiton: theory, methods and applications. Springer, Berlin

130. Bechtold T (2013) Structural topology optimization for MEMS design. University of Freiburg

131. Maheshwaraa U, Seepersad CC, Bourell D (2007) Topology design and freeform fabrication of deployable structures with lattice skins. In: Proceedings of Solid Freeform Fabrication (SFF) Symposium, Austin, TX

132. Fey NP, South BJ, Seepersad CC, Neptune RR (2009) Topology optimization and freeform fabrication framework for developing prosthetic feet. In: Proceedings of Solid Freeform Fabrication (SFF) Symposium, Austin, TX

133. Aremu A, Ashcroft I, Hague R, Wildman R, Tuck C (2010) Suitability of SIMP and BESO topology optimization algorithms for additive manufacture. In: Proceedings of Solid Freeform Fabrication (SFF) Symposium, Austin, TX

134. Biyikli E, To AC (2015) Proportional topology optimization: a new non-gradient method for solving stress constrained and minimum compliance problems and its implementation in MATLAB. Comput Eng Finan Sci 12:e0145041

135. Olason A, Tidman D (2010) Methodology for topology and shape optimization in the design process. Master's Thesis, Chalmers University of Technology, Goteborg, Sweden

136. http://www.3dsystems.com/resources/information-guides/multi-jet-printing/mjp. Accessed June 2016

137. http://www.stratasys.com/3d-printers/technologies/polyjet-technology. Accessed June 2016

138. Gong H, Rafi K, Starr T, Stucker B (2013) The Effects of processing parameters on defect regularity of Ti-6Al-4V fabricated by selective laser melting and electron beam melting. In: Proceedings of Solid Freeform Fabrication (SFF) Symposium, Austin, TX

139. Ponader S, Vairaktaris E, Heinl P, Wilmowsky CV, Rottmair A, Korner C, Singer RF, Holst S, Schlegel KA, Neukam FW, Nkenke E (2008) Effects of topographical surface modifications of electron beam melted Ti-6Al-4V titanium on human fetal osteoblasts. J Biomed Mater Res Part A 84(4):1111–1119

140. Hollander DA, von Walter M, Wirtz T, Sellei R, Schmidt-Rohlfing B, Paar O, Erli H-J (2006) Structural, mechanical and in vitro characterization of individually structured Ti-6Al-4V produced by direct laser forming. Biomaterials 27:955–963

141. Gervasi VR, Stahl DC (2004) Design and fabrication of components with optimized lattice microstructures. In: Proceedings of Solid Freeform Fabrication (SFF) Symposium. Austin, TX

142. Zhang P, Toman J, Yu Y, Biyikli E, Kirca M, Chmielus M, To AC (2015) Efficient design-optimization of variable-density hexagonal cellular structure by additive manufacturing: theory and validation. ASME J Manufact Sci Eng 137(2):021004

143. Gu W (2013) On challenges and solutions of topology optimization for aerospace structural design. In: 10th World Congress on Structural and Multidisciplinary Optimization. Orlando, FL

144. Burns TE (2007) Topology optimization of convection-dominated, steady-state heat transfer problems. Int J Heat Mass Transf 50:2859–2873

145. http://www.industrial-lasers.com/articles/print/volume-29/issue-3/departments/updates/first-metal-3d-printed-bicycle-frame-manufactured.html. Accessed June 2016

146. Gibson LJ, Ashby MF (1997) Cellular solids: structure and properties, 2nd edn. Cambridge University Press, New York

147. Ashby MF, Fleck NA, Gibson LJ, Hutchinson JW, Wadley HNG (2000) Metal forams: a design guide. Butterworth Heinemann, Woburn

148. Banhart J, Weaire D (2002) On the road again: metal foams find favor. Phys Today 55:37–42

149. Banhart J (2006) Metal foams: production and stability. Adv Eng Mater 8:781–794

150. Banhart J (2001) Manufacture, characterization and application of cellular metals and metal foams. Prog Mater Sci 46:559–632

151. Lefebvre L-P, Banhart J, Dunand DC (2008) Porous metals and metallic foams: current status and recent developments. Adv Eng Mater 10(9):775–787

152. http://www.interiorsandsources.com/article-details/articleid/15624/title/alusion-from-cymat-corporation.aspx. Accessed June 2016

153. Vendra LJ, Rabiei A (2007) A study on aluminum-steel composite metal foam processed by casting. Mater Sci Eng, A 465:59–67

154. Jang W-Y, Kyriakides S (2009) On the crushing of aluminum open-cell foams: Part I. Experiments. Int J Solids Struct 469:617–634
155. Tekoglu C (2007) Size effects in cellular solids. Ph.D. Dissertation, University of Groningen, the Netherland
156. Andrews EW, Gioux G, Onck P, Gibson LJ (2001) Size effects in ductile cellular solids. Part II: experimental results. Int J Mech Sci 43:701–713
157. Wang A-J, Kumar RS, McDowell DL (2005) Mechanical behavior of extruded prismatic cellular metals. Mech Adv Mater Struct 12:185–200
158. Ashby MF (2006) The properties of foams and lattices. Philos Trans R Soc A 364:15–30
159. Onck PR, Andrews EW, Gibson LJ (2001) Size effects in ductile cellular solids. Part I: modeling. Int J Mech Sci 43:681–699
160. Wolfram Mathworld. http://mathworld.wolfram.com/Space-FillingPolyhedron.html. Accessed June 2016
161. Murr LE, Gaytan SM, Medina F, Lopez H, Martinez E, Machado BI, Hernandez DH, Martinez L, Lopez MI, Wicker RB, Bracke J (2010) Next-generation biomedical implants using additive manufacturing of complex, cellular and functional mesh arrays. Philos Trans R Soc A 368:1999–2032
162. Yang L (2015) Experimental-assisted design development for an octahedral cellular structure using additive manufacturing. Rapid Prototyping J 21(2):168–176
163. Yang L (2011) Design, structural design, optimization and application of 3D re-entrant auxetic structures. Ph.D. Dissertation, North Carolina State University, Raleigh, NC, USA
164. Yang L, Harrysson O, West H, Cormier D (2015) Shear properties of the re-entrant auxetic structure made via electron beam melting. In: Proceedings of Solid Freeform Fabrication (SFF) Symposium, Austin, TX
165. Bertoldi M, Yardimci MA, Pistor CM, Guceri SI, Danforth SC (1998) Generation of porous structures using fused deposition. In: Proceedings of Solid Freeform Fabrication (SFF) Symposium, Austin, TX
166. Stamp R, Fox P, O'Neill W, Jones E, Sutcliffe C (2009) The development of a scanning strategy for the manufacture of porous biomaterials by selective laser melting. J Mater Sci Mater Med 20:1839–1848
167. Brooks W, Sutcliffe C, Cantwell W, Fox P, Todd J, Mines R (2005) Rapid design and manufacture of ultralight cellular materials. In: Proceedings of Solid Freeform Fabrication (SFF) Symposium, Austin, TX
168. Tsopanos S, Mines RAW, McKown S, Shen Y, Cantwell WJ, Brooks W, Sutcliffe CJ (2010) The influence of processing parameters on the mechanical properties of selectively laser melted stainless steel microlattice structures. J Manuf Sci Eng 132:041011
169. Yang L, Gong H, Dilip S, Stucker B (2014) An investigation of thin feature generation in direct metal laser sintering systems. In: Proceedings of Solid Freeform Fabrication (SFF) Symposium, Austin, TX
170. Harrysson O, Cansizoglu O, Marcellin-Little DJ, Cormier DR, West HA II (2008) Direct metal fabrication of titanium implants with tailored materials and mechanical properties using electron bea melting technology. Mater Sci Eng, C 28:366–373
171. Cansizoglu O (2008) Mesh structures with tailored properties and applications in hips stems. Ph.D. Dissertation, North Carolina State University, Raleigh, NC
172. Yang L, Harrysson O, West II H, Cormier D (2011) Design and characterization of orthotropic re-entrant auxetic structures made via EBM using Ti6Al4V and pure copper. In: Proceedings Solid Freeform Fabrication (SFF) Symposium, Austin, TX
173. Edwards P, Ramulu M (2014) Fatigue performance evaluation of selective laser melted Ti-6Al-4V. Mater Sci Eng, A 598:327–337
174. Gong H, Rafi K, Starr T, Stucker B (2012) Effect of defects on fatigue tests of as-built Ti-6Al-4V parts fabricated by selective laser melting. In: Proceedings of Solid Freeform Fabrication (SFF) Symposium. Austin, TX

175. Murr LE, Gaytan SM, Medina F, Martinez E, Martinez JL, Hernandez DH, Machado BI, Ramirez DA, V RB (2010) Characterization of Ti-6Al-4V open cellular foams fabricated by additive manufacturing using electron beam melting. Mater Sci Eng, A 527:1861–1868
176. Zhang S, Dilip S, Yang L, Miyanaji H, Stucker B (2015) Property evaluation of metal cellular strut structures via powder bed fusion AM. In: Proceedings of Solid Freeform Fabrication (SFF) Symposium. Austin, TX

Chapter 6
The Additive Manufacturing Supply Chain

Abstract The metallic powder bed fusion (PBF) marketplace has been growing rapidly in the second decade of the twenty-first century. The technology has been incorporated into design, engineering, and manufacturing. This chapter will provide insight into the current status of the global 3D printing marketplace with emphasis on additive manufacturing activities and how they are impacting the global supply base. 3D printing will have a major influence on global supply chain logistics and traditional supply chain establishments.

6.1 Production Components Using Metals Powder Bed Fusion

6.1.1 Overview of Current State

Part production using additive manufacturing has been occurring for many years using polymer materials, but metals additive is quickly gaining in this area. For nongovernment regulated components, metalic 3D printing of certain parts is widespread. Examples such as bicycle frames and jewelry are common.

With respect to metallic 3D printing of government regulated parts, the medical industry was one of the first to introduce printed parts into their supply chain environment. Several implant types such as hip and knee joints are now becoming more commonly printed rather than made via traditional manufacturing methods. However, the aerospace industry has been much slower to adopt metals additive technology.

In the commercial aviation market in 2016, General Electric Aviation, MTU/PWA, and Honeywell each have at least one metal powder bed fusion part in the production. While this small number of parts demonstrate great progress, one

© Springer International Publishing AG 2017

L. Yang et al., *Additive Manufacturing of Metals: The Technology, Materials, Design and Production*, Springer Series in Advanced Manufacturing, DOI 10.1007/978-3-319-55128-9_6

has to admit the total number of part numbers in production is relatively low, with estimates of fewer than 10 parts with FAA approvals to date. On the aerospace side, printing components for space and military applications can be somewhat easier because these types of components do not need FAA approvals. Still, the process has a long way to go toward becoming a widespread accepted manufacturing method among aerospace companies.

Aerospace, like medical, is a highly regulated industry where there is not just a cost associated with producing a part, but also a cost from the creation and retaining of manufacturing data. This can manifest itself in several ways, such as the need to track and maintain paperwork on the raw material source, the manufacturing process, and the inspection of these parts.

Traditional manufacturing uses a large supply base with many vendors. The cost to approve and audit suppliers, paperwork and documentation, as well as costs associated with inventory, can loom large—especially for industries that have lower volumes with wide variety of parts. Additive manufacturing offers significant savings with respect to these traditional supply chain pressures and costs. For example, advantages from additive arise when multi-piece components can be built as one-piece systems. When a company is now able to reduce the number of suppliers from several down to one or two, the costs with maintaining manufacturing data and supplier approvals is significantly decreased. These supply chain advantages have helped to propel additive manufacturing technology forward.

Still, as 3D printing technology finds its way into more of the supply base, it offers a financial risk to OEMs if the vendors and the technology are not properly managed. Currently, there are small companies across the globe procuring 3D printing machines and offering to make components for larger OEMs. It should be noted that in most cases, most of the smaller companies can not only not afford the hiring of metallurgists and specialized engineers to oversee the operation of such technology, but most do not understand why they should. The employing of highly educated talent is expensive and the truth is most small companies are not prepared to incur the expense.

From the perspective of an OEM, outsourcing introduces financial risk. Smaller companies offering services will not provide a guarantee that the AM processed material will meet engineering design intent. Smaller companies in the global supply chain may produce a component that meets the engineering needs of the design. However, typically these companies do not have the skill sets required to develop a process that involves selection of the powder, developing the design of experiments necessary to develop the best machine parameters, conducting heat treat trials to evaluate how these steps influence metallographic properties and make a determination if a Hot Isostatic Process (HIP) is required to meet the lifting requirements identified by the design engineer. This is why large OEMs have incorporated vertical integration in their supply chain to minimize risk of component failure once the product is released to the customer.

6.1.2 Steps Toward a Production Process

Getting a production process established begins with the powder company. The OEM must determine how many powder companies will be allowed to bid and supply powder for the additive process. In most cases, to minimize risk of variations in the powder most OEMs will have no more that two powder vendors.

Each powder supplier will begin the selection process by obtaining a quality certification. This might be ISO 9001 or AS9100 (for aerospace). This certification brings a level of confidence to the OEM that sufficient quality controls are in place to insure quality powder is provided on a repeated basis. The OEM will audit this powder company and a "fixed process" will be established and agreed upon. A Fixed Process is a recorded step-by-step document identifying the manufacturing process. Once the fixed process is established, the powder company cannot change any of the steps without approval from the OEM.

Changes to the process can result in significant testing on part of the OEM and changing a fixed process without customer (OEM) approval can result in the powder company being removed from the approved supplier classification. Fixed manufacturing processes are usually developed by the supplier and approved by the OEM cognizant engineering authority.

Adherence to the fixed process is determined via audits conducted on a regular basis by the OEM Supplier Quality Management Group. The fixed process encompasses several aspects of the powder bed fusion process. Documentation of powder certification, how powder is handled and moved into and out of the machine, the machine's printing parameters and all digital files used to build the part are included in the fixed process.

6.1.3 Future Considerations of Metals Additive Manufacturing of Production Parts

While large OEMs are using vertical integration to develop a 3D printing process, industry must recognize this business strategy has its limitations. For large companies with significant volumes of parts requiring additive manufacturing, it is difficult to raise the funding necessary to procure all of the additive manufacturing machines needed to produce their inventory of parts.

During the early phases of production roll-out or Low Rate Initial Production (LRIP), this might be possible. However, at some point OEMs will likely move toward the external supply chain to print production components. This step will require a close working relationship between the OEM and the supply base. What most likely will occur is that OEMs will establish the process internal to their own organization and then once perfected, transfer that process and all the fixed steps associated with that process to an approved supplier.

Over the next few decades, the mechanical and metallurgical properties of metallic parts built with powder bed fusion machines will become common knowledge, and more companies will be comfortable with printing low volume, high-value components. This will bring rise to a growing counterfeit parts market. As a result, infringement on designs and revenue generated from spare part sales will be at risk. One way to avoid infringement will be through the use of unique material chemistries and geometries. Companies will begin to patent these features where possible and protect revenue streams via the granting of licenses.

In the modern global supply chain, the driving focus has been to locate manufacturing in regions of the world that will provide low-cost labor. This action has led to companies focusing on how to ship the components produced in the lowest economical way. However, with 3D printing, the machines, electricity, powder, and argon are all regional or global commodities. Since 3D printing machines run un-attended, the manual labor requirement is significantly reduced. Labor costs are quickly becoming a smaller element of the manufacturing process, and those will force companies to rethink the logic of relocating manufacturing processes.

Today, 3D printing is a small element on the manufacturing and distribution of the supply chain, but 20 years out it will alter decisions made regarding locations and distribution of components.

6.2   Logistics Changes as a Result of 3D Printing

The global logistic value chains are in a state of transformation. Companies like United Parcel Service and Amazon are using 3D polymer printing and altering manufacturing locations to be closer to the end customer. Like the plastic marketplace, the 3D printing of metals will also have an impact on the supply chain. When companies are able to electronically move CAD files across time zones and print the components closer to the end user, then delivery schedules can be shorted and the costs associated with the delivery will be reduced. These cost reductions will cause some companies to rethink the decision to produce components offshore regardless of the country.

Likewise, companies that normally would invent a number of higher margin components will see those inventory levels be reduced. In the coming future, components will be produced when the customer orders the parts. The United States Defense Logistics Agency is an early adopter of this type of strategy, and it is believed other nations and corporations will follow suit with this type of activity.

The movement of goods cost money and time, which in turn may cause a consumer to seek out alternative products that are cheaper and more readily obtainable. This is why logistic companies are using 3D printing to provide products to their customers both faster and with less costs. It should be noted that the real goal of any supply chain is to reduce inventory levels and costs and increase inventory turns. These three things are key elements of any supply chain management effort.

The old traditional supply chain model is focused on low costs and high volumes, which benefit industries such as automotive but often detrimental to aerospace and medical. The concept of print-on-demand means that inventories will not sit in large buildings waiting for an order, which benefits companies in these lower volume, high complexity, and highly regulated industries. Additive manufacturing will enable a new supply chain focus on low volumes, lower inventory, and quick profits. This allows the manufacturing hub to be located anywhere in the world due to most elements being global commodities and influenced very little by labor expenses.

6.2.1 Examples of Additive Manufacturing Transforming the Supply Chain

In 2016, the U.S. Navy installed a 3D printer on the amphibious assault ship, USS Essex, for testing. The Navy is training sailors to handle this technology, so if there is a part needed, and it does not exist in the onboard inventory, sailors can design and print the part on demand within hours or days, allowing for a more rapid response to the ship's needs.

Shipping giant, Maersk, is looking to experiment with 3D printing, hoping to save costs on current high transportation expenses. The idea is to install a 3D printer on a cargo vessel to allow the crew to print spare parts on demand. Maersk will send a blueprint to the crew, and they will simply push "print," and in a matter of hours they will have the required part.

The shipping giant Amazon filed for a patent for 3D printers mounted within trucks, which could then print customers' purchases on the fly and deliver them instantly. The business case for this idea is that printing the component near the user will save money by removing the necessity to import the part and/or store them in a warehouse.

Another company transcending supply chain delivery rules is CloudDDM, a supplier of additive manufacturing services for various industries. Cloud DDM and their hundreds of printing machines are co-located on a UPS shipping hub. By having such close proximity to shipping services, the company is able to print customer parts and have them shipped overnight provided they can be printed by the cut-off time. Compared to other additive manufacturers, being co-located on an UPS shipping hub campus allows CloudDDM an additional 6 h of printing time before shipping cutoffs.

Consulting companies are now predicting that 3D printing will have a major impact on the global supply chain. In 2015, per consulting firm Frost & Sullivan, 3D printing will see an increase in market share, possibly as much as 40% over a span of 12 months and will continue to increase in the foreseeable future. This new technology will have an influential impact on the supply chain and play a role in its global disruption for the foreseeable future.

6.3   The Next 20 Years—Where the Metal 3D Printing Supply Base Is Headed

6.3.1   From the Perspective of Components and Production

In 20 years, 3D printing of components will be a commonly accepted manufacturing method. Since the majority of corporations will not be able to directly procure—either due to cost or space—the machines needed to print large portions of their inventory, the supply base for metal additively manufactured parts will increase. This means the mostly vertically integrated supply chain of today will transform into a more horizontal supply chain.

Over the next two decades, it is projected that the material properties of 3D printed will be more readily available. Some of this data will be generated by government agencies and made accessible to the public. Some of this data will be generated by large corporations and will be retained as Intellectual Property. These corporations may license this engineering data to their supply chains with the intent of designing and building components.

6.3.2   From the Perspective of Logistics

There has been no shortage of news stories reporting on how militaries across the globe all seem to suffer from the same calamity, which is a broken supply chain and its inability to supply components on a timely manner at a reasonable cost. However, the next two decades will document a major transformation.

It is happening now in some instances but 20 years from now, the Armed Services will use 3D Printing Technology to resolve major logistics issues. One major logistics problem almost all military and major corporations alike face is the length of the supply chain to procure critical parts. It is realistic to envision an aircraft carrier with 40–100 machines and a full machine shop and heat treat facility to print metal components rather than procure them using conventional manufacturing methods.

By locating the means of manufacture next to the customer, the supply chain will be considerably shortened and the requirements for large inventories will be reduced. As metals 3D printing becomes more accepted and widespread within the industry, the companies and organizations that are quick to adapt their supply chain toward this new technology will thrive compared to any competition that continues to rely on the traditional supply chain strategy.

References

1. http://info.plslogistics.com/blog/6-effects-3d-printing-has-on-supply-chains; PLS Logistics, 16 Apr 2015
2. 3D Printing Central to Nike's Continued Growth According to CEO Parker; Ralph Barnette, 9 Oct 2014
3. The UPS store expands 3D printing across the U.S.A.; 24/7 Staff, 1 Oct 2014
4. http://www.supplychain247.com/search/results/search&keywords=3D+Printing/. Supply chains and third-party logistics make room for 3D printing; Georgia Tech Manufacturing Institute, 23 Jan 2014
5. Amazon launches 3D printing services; John Newman, 30 Jul 2014
6. Amazon & Home Depot to Sell 3D Printing; 24/7 Staff, 23 Sept 2014
7. The implications of 3D printing for the global logistics industry; John Manners-Bell & Ken Lyon, 23 Jan 2014
8. http://cerasis.com/2014/02/10/3d-printing-supply-chain/, 13 Feb 2013
9. 3D Systems Teams Up with the White House to Transform American Manufacturing; 13 Mar 2014
10. What should be on the supply chain radar? Predictions for 2014; Irfan Khan, 16 Feb 2014
11. Why are we letting digital marketers define the future world view of the supply chain? Lora Cecere, 30 Jul 2015
12. Digital Supply Chain—Insights on Driving the Digital Supply Chain Transformation; Lora Cecere, 22 Jan 15
13. The supply chain is dead; Long live the value web; Cognizant, 3 Nov 2014
14. Collaboration with industry shapes 25 years of Center for Supply Chain Research; 24/7 Staff, 8 June 2014
15. Advanced supply chain capabilities are a crucial catalyst for strong financial performance; Deloitte, 15 Apr 2014
16. Talent is the future of supply chain management: 24 Jan 2014 at MIT's Research Expo
17. Talent is the future of supply chain management; MIT; 23 Jan 2014
18. Omni-channel and supply chain analytics top trends in Deloitte MHI industry report; Roberto Michel, Editor at Large, 11 Apr 2014
19. Balancing priorities in the supply chain; Qlik, 28 Jul 2015
20. Supply chain talent of the future survey; Kelly Marchese & Ben Dollar, 5 May 2015
21. MHI Annual Industry Report: Supply Chain Innovation—Making the Impossible Possible; Deloitte & MHI
22. Is now the time to try direct digital manufacturing? Scott Crump, 7 Oct 2014
23. Modex 2014: Three keynotes to examine trends, share insights; 16 Jan 2014
24. Innovative or inconclusive? Evaluating new supply chain ideas; Ken Cottrill & Jim Rice, 4 May 2015
25. Transforming the future of supply chains through disruptive innovation; Ken Cottrill, 30 Jul 2014
26. Transforming the future of supply chains through disruptive innovation, 8 Nov 2011
27. Strategic sourcing: time for an integrated transportation strategy, 1 June 2015
28. What could 3D printing mean for the supply chain? Andrew Bell, 1 Jan 2014
29. 3D printing and the supply chains of the future; Mark Patterson, 1 Jan 2014
30. Amazon files patent for 3D printing delivery trucks; Brian Krassenstein, 4 Mar 15
31. MHI & Deloitte Study: Traditional supply chains to undergo radical transformation by 2025; Deloitte & MHI, 30 Mar 2015
32. 3D printing operation on UPS supply chain campus; 24/7 Staff, 7 May 2015
33. 3D printing promises to transform manufacturing and supply chains; Jeremy Thomas, 9 Oct 2014
34. Once peripheral technologies take center stage in new Freedonia research, 26 Feb 2014

35. Your current supply chain is toast, but its replacement will be awesome; Paul Brody, 24 June 2013
36. Trends that will shape the supply chain in 2014; Supply Chain Matters—Mark Patterson, 28 Jan 2014
37. 3D systems celebrates manufacturing day with open house events Nationwide, 3 Oct 2013
38. Fifth-annual UPS Change in the (Supply) Chain survey takes deep dive into high-tech supply chains, 21 May 2015. http://www.roboticstomorrow.com/article/2014/07/the-impact-of-3-d-printing-on-supply-chains/4373/
39. Advanced Manufacturing, Business Development; Len Calderon, Robotics Tomorrow, 24 Jul 2014
40. The impact of 3D printing in the supply chain and logistics arenas; 3D printing logistics supply chain, Chuck Intrieri, 10 Feb 2014. https://www.loaddelivered.com/blog/how-will-3d-printing-impact-supply-chain/
41. Amazon files patent for 3D printing delivery trucks; Brian Krassenstein, 4 Mar 2015
42. MHI & Deloitte Study: Traditional supply chains to undergo radical transformation by 2025; Deloitte & MHI, 30 Mar 2015
43. 3D printing operation on UPS supply chain campus; 24/7 Staff, 7 May 15. https://3dprint.com/62642/cloudddm-ups/

Printed in the United States
By Bookmasters